PHYSICS RESEARCH AND TECHNOLOGY

OPTICALLY INDUCED SPACE-CHARGE GRATINGS IN WIDE-BANDGAP SEMICONDUCTORS

TECHNIQUES AND APPLICATIONS

PHYSICS RESEARCH AND TECHNOLOGY

Additional books in this series can be found on Nova's website under the Series tab.

Additional E-books in this series can be found on Nova's website under the E-book tab.

PHYSICS RESEARCH AND TECHNOLOGY

OPTICALLY INDUCED SPACE-CHARGE GRATINGS IN WIDE-BANDGAP SEMICONDUCTORS

TECHNIQUES AND APPLICATIONS

IGOR A. SOKOLOV
AND
MIKHAIL A. BRYUSHININ

nova
science publishers
New York

For permission to use material from this book please contact us:
Telephone 631-231-7269; Fax 631-231-8175
Web Site: http://www.novapublishers.com

Library of Congress Cataloging-in-Publication Data

Names: Sokolov, Igor A. (Scientist in physics), author. | Bryushinin, Mikhail A., author.
Title: Optically induced space-charge gratings in wide-bandgap semiconductors :techniques and applications / Igor A. Sokolov and Mikhail A. Bryushinin (Ioffe Institute, St.-Petersburg, Russia).
Description: Hauppauge, New York: Nova Science Publishers, Inc., [2017] | Series: Physics research and technology | Includes bibliographical references and index.
Identifiers: LCCN 2017019931 (print) | LCCN 2017023801 (ebook) | ISBN 9781536119640 | ISBN 9781536119442 | ISBN 9781536119442(hardcover) | ISBN 1536119442 (hardcover) | ISBN 9781536119640(ebook) | ISBN 1536119644(ebook)
Subjects: LCSH: Nonlinear optics. | Photorefractive materials. | Electromotive force. | Wide gap semiconductors.
Classification: LCC QC446.2 (ebook) | LCC QC446.2 .S65 2017 (print) | DDC 537.6/226--dc23
LC record available at https://lccn.loc.gov/2017019931

Published by Nova Science Publishers, Inc. † New York

Contents

Preface

The photorefractive effect is one of the highest-sensitivity non-linear optical effects. The rate of progress in the development of photorefractive applications has been limited by the rate in which breakthroughs in materials science have supplied better wide-bandgap semiconductor materials. The last years have seen an upsurge of interest in characterization of new wide-bandgap semiconductor materials. Holographic recording in wide-bandgap photorefractive materials is of particular interest from the scientific and practical point of view. Such a recording includes the stage of space charge formation, which is associated with the spatial redistribution of photoinduced carriers. A number of new methods studying the dynamics of space charge and photoconductivity in wide-bandgap semiconductors were proposed during recent years.

This book deals with the so-called non-steady-state photo-EMF effect and its applications for material characterization and adaptive interferometry. The first part of the book represents theory of the non-steady-state photo-EMF effect. The second part of the book describes application of the non-steady-state photo-EMF effect for investigation of wide-bandgap semiconductors and nanostructured materials including photorefractive sillenites grown in the air and argon atmosphere, molecular crystal SnS_2, boron nitride crystal, nanosized GaN and Se within porous matrices, SiC, AlN and Ga_2O_3 crystals. The third part of the book presents theoretical analysis and experimental investigation of the non-steady-state photoelectromotive force and two-wave mixing in photorefractive crystals under frequency-modulated illumination. The final part of the book describes adaptive interferometers using the effect of the non-steady-state photo-EMF. This book has been written to be accessible to the specialists working in solid state physics, semiconductor physics, adaptive interferometry, laser-based ultrasound. The subject areas include: photorefractive non-linear optics, dynamic holography, interferometry, wide-bandgap semiconductors, laser-doppler velocimetry. References to over 200 original papers represent the personal selection of the authors. We are grateful to all our colleagues for their assistance and contributions.

The authors acknowledge the financial support from the Russian Science Foundation (grant 15-12-00027).

Igor A. Sokolov
Mikhail A. Bryushinin
Ioffe Institute, St.-Petersburg, Russia
February 2017

Introduction

The process of light interaction with condensed matter is an important problem of the modern solid state physics and quantum electronics [1]- [5]. Holographic recording in wide-bandgap semiconductors and nano-structured materials is of particular interest from the scientific and practical point of view. Practical applications are associated with the detection of low-amplitude acoustic vibrations of real objects, such as ultrasonic transducers, composite materials, biological objects, metal production, MEMS [6]- [13]. Conventional techniques using homodyne laser vibrometers [7] are suitable for practical applications in this regard for their high sensitivity which is limited in principle only by shot-noise of the laser used. Their utilization is restrained by several problems such as slow phase drifts in the interferometer arms due to environmental reasons, necessity of fine optical adjustment and suppression of laser amplitude noise. Photorefractive wide-bandgap semiconductor materials offer a novel elegant way to solve the problem of keeping operation interferometer point constant [2, 14–17]. The crystal replaces the conventional beam-splitter and can be controlled not only electrically but also optically, i.e., based on the principles of nonlinear optics. In addition, it can be multilayered. Such a multilayered adaptive beam-splitter is nothing else but a volume dynamic hologram recorded by the reference and signal waves.

Holographic recording includes the stage of space charge formation, which is associated with the spatial redistribution of photoinduced carriers [18]. The nonstationary holographic photocurrent [19] (or non-steady-state photo-EMF [20]) is a holography related effect and it appears in a semiconductor material illuminated by an oscillating light pattern. Such illumination is usually created by two coherent light beams one of which is phase modulated with frequency

ω. The alternating current is resulted from the periodic relative shifts of the photoconductivity and space charge gratings which arise in the crystal's volume under illumination. Like the holographic recording in photorefractive crystals this effect demonstrates adaptive properties that promote its application in such areas as vibration monitoring, velocimetry, etc. [21]. The sensitivity of adaptive photodetector is comparable with the sensitivity of classical schemes using photodiodes. The amplitude of the non-steady-state photo-EMF can be increased by an external ac field so one can expect the improvement of the sensitivity in certain ranges of temporal and spatial frequencies. In contrast to the holographic methods this technique allows the direct transformation of phase modulated optical signals into the electrical current and can be applied for characterization of centrosymmetrical and even amorphous materials [23]- [32]. Since the photocurrent is originated from the interaction of both the photoconductivity and space charge gratings a lot of photoelectric parameters can be measured [18,20].

Photorefractive sillenites $Bi_{12}Si(Ge,Ti)O_{20}$ are promising materials for optical signal processing due to their high sensitivity [2]. Large electrooptic coefficient allows an easy observation of the internal electric field. Shallow traps play significant role in the process of hologram formation. Recording and erasing processes have several time constants [33, 34]. Many attempts have been started in order to determine the charge carrier mobility, which is among the most important parameters of the light-induced charge transport. For photorefractive sillenites electron mobility values measured by various techniques from 5.2×10^{-9} to 3.4×10^{-4} $m^2V^{-1}s^{-1}$ have been reported [35–39]. This scattering of the data is usually attributed to the influence of shallow traps which reduce the average velocity of excited carriers. The first part of the book represents theory of the non-steady-state photo-EMF effect. The second part of the book describes application of the non-steady-state photo-EMF effect for investigation of wide-bandgap semiconductors and nanostructured materials including photorefractive sillenites grown in the air and argon atmosphere, molecular crystal SnS_2, boron nitride crystal, nanosized GaN and Se within porous matrices, SiC, AlN and Ga_2O_3 crystals. The third part of the book presents theoretical analysis and experimental investigation of the non-steady-state photoelectromotive force and two-wave mixing in photorefractive crystals under frequency-modulated illumination. The final part of the book describes adaptive interferometers using the the effect of the non-steady-state photo-EMF.

Chapter 1

Non-Steady-State Photo-EMF in Semiconductors with Shallow Traps

1. Diffusion Regime

1.1. Electric Field and Photocurrent Amplitudes

This section presents calculations of the non-steady-state photo-EMF amplitude for the simplest two-level model of semiconductor with electron photoconductivity [22, 40]. Photocarriers are excited from the deep donor levels (with the total density N_D) and from the shallow levels (with the total density M_T) [Fig. 1.1(a)]. The deep donors are ionized by light radiation only, for the shallow traps both the light and thermal mechanisms of activation are possible. The shallow level is assumed to be empty in the dark and the deep donor level is partially filled (with the density of vacant states N_A). The balance equations for this model can be written as follows [41]:

$$\frac{\partial N}{\partial t} = S_D I(N_D - N_A - N) - \gamma_D n(N_A + N), \tag{1.1}$$

$$\frac{\partial M}{\partial t} = -(\beta + S_T I)M + \gamma_T n(M_T - M). \tag{1.2}$$

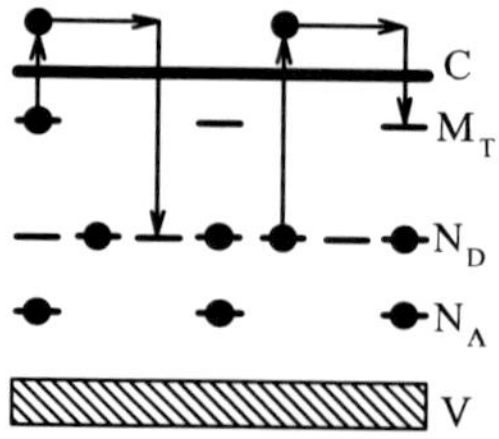

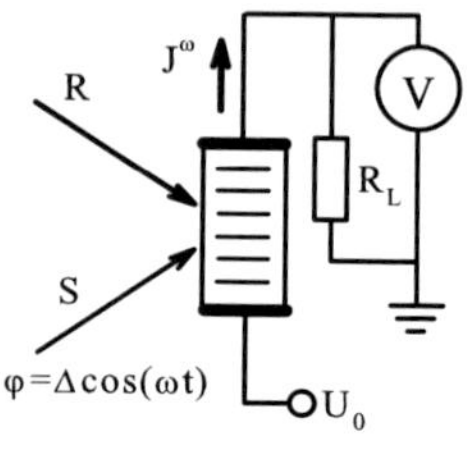

Figure 1.1. Model of a semiconductor with shallow levels (left figure) and the conventional scheme of holographic photocurrent measurements (right figure).

Here N is the density of deep centers emptied by light ($N_D^+ = N_A + N$ is the total concentration of ionized donors), M is the density of shallow centers filled with electrons, n is the density of free electrons, S_D and S_T are the light excitation cross sections for deep and shallow levels, β is the thermal excitation rate from shallow level, γ_D and γ_T are the recombination constants for deep and shallow traps respectively, I is the light intensity.

For calculation of the non-steady state photocurrent amplitude kinetic equations (1.1), (1.2) should be added by the continuity and Poisson's equations and by the expression for the current density:

$$\frac{\partial}{\partial t}(N - M - n) + \frac{1}{e}\operatorname{div}\mathbf{j} = 0, \tag{1.3}$$

$$\operatorname{div}\mathbf{E} = \frac{e}{\epsilon\epsilon_0}(N - M - n), \tag{1.4}$$

$$\mathbf{j} = e\mu n\mathbf{E} + eD\operatorname{grad} n. \tag{1.5}$$

Here $\mathbf{j}$ is the current density, $\mathbf{E}$ is the electric field, μ is the electron mobility, $D = (k_B T/e)\mu$ is the diffusion coefficient, ϵ is the dielectric constant of the medium, ϵ_0 is the permittivity of free space, e is the electron charge.

Let us consider the excitation of the non-steady-state photo-EMF with an oscillating interference pattern formed by two coherent plane waves one of which is phase modulated with frequency ω and amplitude Δ [Fig. 1.1(b)]:

$$I(x,t) = I_0\{1 + m\cos[Kx + \Delta\cos(\omega t)]\}. \tag{1.6}$$

Here I_0 is the average light intensity, m is the contrast and K is the spatial frequency of the interference pattern [20]. For small contrast of the interference pattern m and small amplitude of phase modulation Δ ($m \ll 1$, $\Delta \ll 1$) the intensity distribution (1.6) can be written as [42]:

$$\begin{aligned} I(x,t) &= I_0 + I^{+0}\exp(iKx) + I^{-0}\exp(-iKx) \\ &\quad + I^{++}\exp[i(Kx+\omega t)] + I^{+-}\exp[i(Kx-\omega t)] \\ &\quad + I^{-+}\exp[i(-Kx+\omega t)] + I^{--}\exp[i(-Kx-\omega t)], \end{aligned} \tag{1.7}$$

where

$$\begin{aligned} I^{+0} &= I^{-0} = \frac{m}{2}I_0, \\ I^{++} &= I^{+-} = -I^{-+} = -I^{--} = \frac{im\Delta}{4}I_0. \end{aligned} \tag{1.8}$$

We shall look for the solution of Equations (1.1) - (1.5) in the form of identical decompositions for $N(x,t)$, $M(x,t)$, $n(x,t)$, $E(x,t)$ as it was done for the intensity distribution. All the necessary complex amplitudes for N, M, n and E have been found and the most important of them can be written as follows:

$$n^{\pm 0} = \frac{m}{2}\nu_0, \tag{1.9}$$

$$n^{\pm +} = \frac{\pm im\Delta}{4}\nu_1, \tag{1.10}$$

$$E^{\pm 0} = \frac{\mp im}{2}(\nu_0/n_0)E_d, \tag{1.11}$$

$$E^{\pm +} = \frac{m\Delta}{4}\frac{\nu_1/n_0}{1+i\omega\tau_M}E_d, \tag{1.12}$$

where

$$\nu_0 = \frac{S_D I_0(N_D - N_A - N_0)T_D + S_T I_0 M_0 T_T}{1 + T_D/\tau_D + T_T/\tau_T + \tau_M D K^2}, \tag{1.13}$$

$$\nu_1 = \frac{S_D I_0(N_D - N_A - N_0)T_D/(1+i\omega T_D) + S_T I_0 M_0 T_T/(1+i\omega T_T)}{1 + T_D/[\tau_D(1+i\omega T_D)] + T_T/[\tau_T(1+i\omega T_T)] + \tau_M D K^2/(1+i\omega\tau_M)}, \tag{1.14}$$

N_0, M_0, n_0 are the stationary and spatially uniform concentrations, $E_d = (k_B T/e)K$ is the diffusion field, $\tau_M = \epsilon\epsilon_0/\sigma_0$ is the Maxwell relaxation time, $\sigma_0 = e\mu n_0$ is the average photoconductivity,

$$\begin{aligned} T_D &= (S_D I_0 + \gamma_D n_0)^{-1}, \\ T_T &= (\beta + S_T I_0 + \gamma_T n_0)^{-1} \end{aligned} \tag{1.15}$$

are the inverse values of the sum of the generation and ion recombination rates for deep and shallow levels [43],

$$\begin{aligned} \tau_D &= [\gamma_D(N_A + N_0)]^{-1}, \\ \tau_T &= [\gamma_T(M_T - M_0)]^{-1} \end{aligned} \tag{1.16}$$

are the recombination times of the electrons to the deep and shallow traps. Note, that the obtained relation for the electric field (1.12) can also be useful for the analysis of two-wave mixing in photorefractive materials.

The total photocurrent through the short-circuited sample (for cyclic boundary conditions) is defined by the drift component of current density averaged over the interelectrode spacing L [20]:

$$j(t) = \frac{1}{L}\int_0^L e\mu n(x,t)E(x,t)dx. \tag{1.17}$$

As follows from Eq. (1.17) the contribution to the first harmonic (with frequency ω) of the non-steady-state photocurrent density amplitude is produced only by the following combination of coefficients:

$$j^\omega = 2e\mu(n^{++}E^{-0} + n^{-+}E^{+0} + n^{+0}E^{-+} + n^{-0}E^{++}). \tag{1.18}$$

Finally, for the complex amplitude of the current density we obtain the following expression:

$$j^\omega = \frac{m^2\Delta}{2}\sigma_0 E_d \frac{-i\omega\tau_M}{1 + i\omega\tau_M}\frac{\nu_0\nu_1}{n_0^2}. \tag{1.19}$$

It is important to point out that all expressions for the electric field and photocurrent amplitude were obtained with the only approximation of small contrast m and amplitude of phase modulation Δ (no assumptions were made about the occupancy of the deep and shallow levels and the prevalence of any activation mechanism for the shallow level).

1.2. Low Intensity Approximation

The expression for the complex amplitude of the photo-EMF [Eq. (1.19)] contains stationary and spatially uniform concentrations N_0, M_0 and n_0 that appear in the sample under uniform illumination with the average intensity I_0. The values N_0, M_0 and n_0 are calculated from (1.1), (1.2) and the neutrality condition:

$$N_0 = M_0 + n_0. \tag{1.20}$$

In this section we consider the situation of low illumination levels, i.e., slight filling of the shallow level:

$$M_0 \ll M_T. \tag{1.21}$$

The strong filling of the shallow level makes the effective mobility and lifetime tend to their real values [44], this situation is considered separately in section 1.3..

The simplest solution of equations (1.1), (1.2), (1.20) can be obtained in the case of linear generation from the deep level

$$N_0 \ll N_D - N_A \tag{1.22}$$

and predominance of the thermal activation of shallow traps under light excitation

$$S_T I_0 \ll \beta. \tag{1.23}$$

The assumption (1.22) means that we neglect light-induced changes in absorption [41]. If conditions (1.21) - (1.23) are satisfied the concentrations N_0, n_0 equal

$$N_0 = \sqrt{\frac{N_A^2}{4} + \frac{S_D I_0 (N_D - N_A)}{\gamma_D}\left(1 + \frac{M_T}{N_{CM}}\right)} - \frac{N_A}{2}, \tag{1.24}$$

$$n_0 = \frac{N_0}{(1 + M_T/N_{CM})}, \tag{1.25}$$

where

$$N_{CM} = \beta/\gamma_T = N_C \exp[-E_T/(k_B T)] \tag{1.26}$$

is the effective density of states in the conduction band calculated with respect to the energy position of shallow traps [44], N_C is the effective density of states in the conduction band, E_T is the energy depth of the shallow level, k_B is the Boltzmann constant, T is the absolute temperature.

The concepts of real and effective photoelectric parameters were considered elsewhere [33,44]. For the considered model, the relation between effective and real parameters is the following:

$$\mu/\mu' = \gamma_D/\gamma'_D = (1 + M_T/N_{CM}). \tag{1.27}$$

This expression denotes that all the values containing N_0 (e.g., Debye screening length L_S), μn_0-product (e.g., average photoconductivity σ_0) and $\mu\gamma_D^{-1}$-product (e.g., diffusion length L_D) can be written using either real or effective parameters.

There are two significant types of intensity dependencies of n_0 and N_D^+ (and as a consequence, photoconductivity and effective lifetime of carriers). If the intensity of the incident radiation is small enough, so that $N_0 \ll N_A$, the effective lifetime $\tau' = (\gamma'_D N_D^+)^{-1}$ is independent on light intensity and photoconductivity is the linear function of light intensity ($\sigma_0 \propto I_0$). For the opposite situation, i.e., for $N_0 \gg N_A$ (the case of quadratic recombination), the quantities σ_0 and τ' depend on the intensity as: $\sigma_0, 1/\tau' \propto I_0^{0.5}$.

As seen the general expression (1.19) is rather complicated for further analysis, so we shall consider the frequency transfer function of the photo-EMF under certain simplifying assumptions. Three of them, i.e., Eqs. (1.21) - (1.23), were made when we calculated the spatially uniform concentrations. The fourth condition can be written as follows:

$$S_T M_0 \ll S_D(N_D - N_A). \tag{1.28}$$

This is the case when the light excitation of free carriers from the deep level is more effective than the light excitation from the shallow one. Another condition can be written as

$$\beta T_D \gg 1. \tag{1.29}$$

This condition states that the number of free carriers excited from the shallow level during characteristic time T_D is larger than the number of carriers captured at this level (for $N_D \gg N_A + N_0$ the value T_D is nearly equal to the lifetime of ionized deep level, i.e., $T_D \approx (\gamma_D n_0)^{-1}$). Under these simplifying assumptions the expression for the photocurrent amplitude is the following:

$$j^{\omega} = \frac{-0.5m^2\Delta\sigma_0 E_d(1+\Theta+K^2L_S^2)^{-1} i\omega\tau_M}{1+\Theta+K^2L_S^2-\omega^2\tau_M\tau' + i\omega[\tau'+\tau_M(1+\Theta+K^2L_D^2)]}. \quad (1.30)$$

Here

$$L_D = \sqrt{\mu\tau_D k_B T/e} \quad (1.31)$$

is the average diffusion length of photocarriers (corresponding to deep traps),

$$\tau' = \tau'(\omega) = \tau_D\left(1+\frac{M_T/N_{CM}}{1+i\omega\beta^{-1}}\right) \quad (1.32)$$

is complex and frequency-dependent effective electron lifetime,

$$L_S = \sqrt{\frac{\epsilon\epsilon_0 k_B T}{e^2(N_A+N_0)(1-N_A/N_D)}} \quad (1.33)$$

is the Debye screening length [41,42],

$$\Theta = [(1+N_A/N_0)(1-N_A/N_D)]^{-1} \quad (1.34)$$

is the dimensionless parameter ($0 < \Theta < 1$). This parameter characterizes the type of captures to the deep level: $\Theta \ll 1$ for linear recombination ($N_0 \ll N_A$, $\sigma_0 \propto I_0$) and Θ tends to 1 for quadratic recombination ($N_0 \gg N_A$, $\sigma_0 \propto I_0^{0.5}$).

For the numerical calculations of the photo-EMF we use the following parameters [2,33,39,45]:

$$\begin{aligned}
&N_D = 1\times10^{25}\ \text{m}^{-3}, && N_A = 0.95\times10^{22}\ \text{m}^{-3},\\
&M_T = 10^{20}-10^{21}\ \text{m}^{-3}, && \mu = 3.4\times10^{-4}\ \text{m}^2\text{V}^{-1}\text{s}^{-1},\\
&S_D = 1.06\times10^{-5}\ \text{m}^2\text{J}^{-1}, && \beta^{-1} = 2\times10^{-7}\ \text{s},\\
&\gamma_D' = 1.65\times10^{-17}\ \text{m}^3\text{s}^{-1}, && \tau_T = 4\times10^{-9}\ \text{s}.
\end{aligned} \quad (1.35)$$

Conditions (1.21) - (1.23), (1.28), (1.29) are satisfied for moderate light intensities $I_0 = 0 - 3\times10^4$ W/m^2. It is useful to write the values contained in (1.30) and calculated for $I_0 = 10^4$ W/m^2:

$$\begin{aligned}
\tau_D &= 1.3\times10^{-7}\ \text{s}, & \tau'(0) &= 6.4\times10^{-6}\ \text{s},\\
\tau_M &= 6.9\times10^{-5}\ \text{s}, & T_D &= 9.0\times10^{-3}\ \text{s},\\
T_T &= 2.0\times10^{-7}\ \text{s}, & L_D &= 1.0\times10^{-6}\ \text{m},\\
L_S &= 9.1\times10^{-8}\ \text{m}, & \Theta &= 7.1\times10^{-4}.
\end{aligned} \quad (36)$$

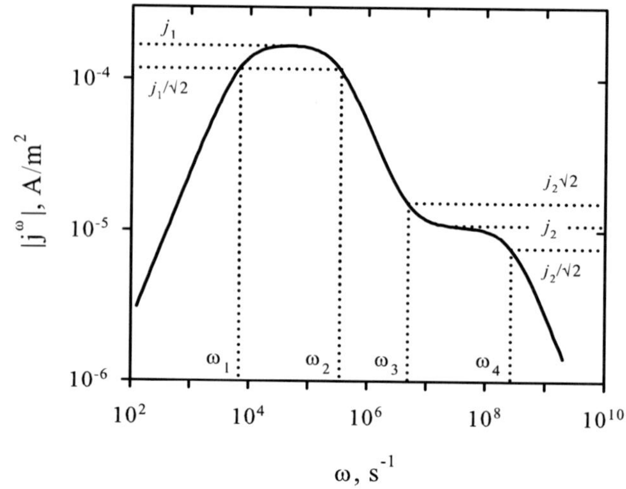

Figure 1.2. Theoretical dependence of the photocurrent amplitude j^ω versus frequency of phase modulation ω ($I_0 = 10^4$ W/m^2, $K = 10^6$ m^{-1}, $m = 0.2$, $\Delta = 0.1$).

Note that $T_T = \beta^{-1}$ if assumptions (1.21), (1.23) take place. As seen conditions $\tau_T \ll \beta^{-1} \sim \tau_D \ll \tau'(0) \ll \tau_M \ll T_D$, $L_S \ll L_D$, $\Theta \ll 1$ are satisfied.

The frequency dependencies of the photocurrent amplitude and its phase are shown in Figs. 1.2, 1.3. The curve $|j^\omega(\omega)|$ has one growing, two frequency independent and two decreasing regions.

The linear growth at low frequencies following by the frequency independent region has the standard explanation [20,42]. For low oscillation frequencies ω both the photoconductivity and electric field gratings follow the movements of the interference pattern. This results in a small value of periodical spatial shifts between these distributions, and as a consequence, in a small amplitude of the output electrical signal.

In order to clarify this situation let us consider the frequency dependencies of complex amplitudes n^{-0}, n^{++}, E^{-0}, E^{++} (Fig. 1.4) and estimate their relative contribution to the total current (1.18). As seen from Fig. 1.4(a,b) for $\omega \ll \omega_1$ the following relation between corresponding complex amplitudes n^{-0}, n^{++}, E^{-0}, E^{++} is valid: $|n^{++}E^{-0}| \approx |n^{-0}E^{++}|$ (or $|n^{-0}/n^{++}| \approx$

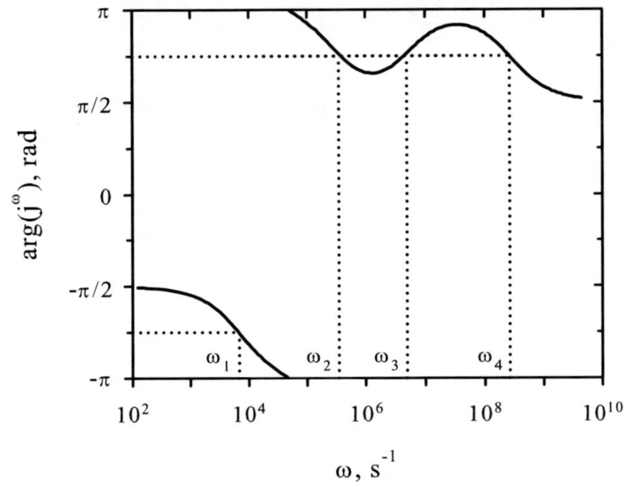

Figure 1.3. Frequency dependence of the photocurrent phase ($I_0 = 10^4$ W/m^2, $K = 10^6$ m^{-1}, $m = 0.2$, $\Delta = 0.1$).

$|E^{-0}/E^{++}|$). However the phases of products $n^{++}E^{-0}$ and $n^{-0}E^{++}$ are opposite [Fig. 1.4(c)] and these components compensate each other resulting in a small photocurrent amplitude. If the frequency of phase modulation is increased the oscillation amplitudes n^{++} and E^{++} become smaller with the faster decrease of the latter component [Fig. 1.4(a,b)]. Besides an additional phase shift appears between them [Fig. 1.4(c)]. Finally we have $n^{++}E^{-0} + n^{-0}E^{++} \propto -i\omega\tau_M$ for $\omega < \omega_1$. The first cut-off frequency ω_1 is defined as

$$\omega_1 \approx \frac{1 + \Theta + K^2 L_S^2}{\tau'(0) + \tau_M(1 + \Theta + K^2 L_D^2)}. \tag{1.37}$$

For the "relaxation type" photoconductor ($\tau'(0) \ll \tau_M$) and for oscillation frequencies $\omega > \tau_M^{-1}$ the electric field grating can be considered as "frozen in" (i.e., $|n^{++}E^{-0}| > |n^{-0}E^{++}|$), the amplitude of the photocurrent reaches its maximum $|j^\omega| \propto |n^{++}E^{-0}|$ and becomes frequency independent up to the

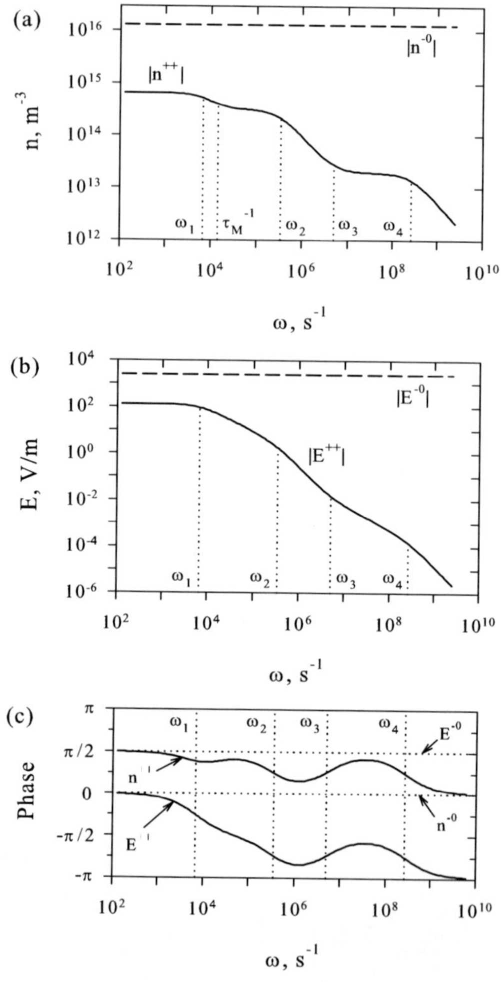

Figure 1.4. Amplitudes of electron concentration (a), electric field (b) and their phases (c) versus frequency of phase modulation ω ($I_0 = 10^4$ W/m^2, $K = 10^6$ m^{-1}, $m = 0.2$, $\Delta = 0.1$).

second characteristic frequency ω_2 [Figs. 1.2, 1.4(a)]:

$$\omega_2 \quad \approx \quad \frac{1}{\tau_M} + \frac{1 + \Theta + K^2 L_D^2}{\tau'(0)}$$

$$+\frac{1+\Theta+K^2L_S^2}{\tau'(0)+\tau_M(1+\Theta+K^2L_D^2)}. \tag{1.38}$$

In the "lifetime" regime ($\tau'(0) \gg \tau_M$) relation $|n^{++}E^{-0}| \approx |n^{-0}E^{++}|$ maintain for $\omega < \omega_2 \approx \tau_M^{-1}$ and dependence $n^{++}E^{-0} + n^{-0}E^{++} \propto -i\omega\tau_M$ is expected but both complex amplitudes n^{++} and E^{++} decrease as $1/\omega$ for $\omega > \omega_1 \approx \tau'(0)^{-1}$ giving a plateau between ω_1 and ω_2.

The photocurrent amplitude for the first plateau for both "relaxation" and "lifetime" regimes can be written as follows:

$$j_1 \approx \frac{-0.5m^2\Delta\sigma_0 E_d(1+\Theta+K^2L_S^2)^{-1}}{1+\Theta+\tau'(0)/\tau_M+K^2L_D^2}. \tag{1.39}$$

The photo-EMF signal peaks at modulation frequency equal to

$$\omega_{m1} \approx \sqrt{\frac{1+\Theta+K^2L_S^2}{\tau_M\tau'(0)}}. \tag{1.40}$$

For high excitation frequencies $\omega > \omega_2$ the photocurrent amplitude falls down because of corresponding decrease of the photoconductivity grating amplitude n^{++} [Fig. 1.4(a)]. There are two regions on the frequency dependence of photocurrent where the signal amplitude decreases inversely proportional to the frequency of phase modulation ($\propto 1/\omega$). This behavior is associated with the frequency dependence of the effective electron lifetime τ'. As seen from Eq. (1.32) there are two frequency regions where the effective lifetime τ' is real and frequency-independent:

$$\tau' \approx \tau'(0) = \tau_D(1 + M_T/N_{CM}) \tag{1.41}$$

for excitation frequencies much smaller than the thermal excitation rate from the shallow level ($\omega \ll \beta$) and

$$\tau' \approx \tau_D \tag{1.42}$$

for high excitation frequencies $\omega \gg \tau_T^{-1}$. The parameter $\tau'(0)$ was introduced earlier in Ref. [44] as a time constant describing the process of photoconductivity relaxation in semiconductor with shallow traps. When the crystal is illuminated with square modulated light the photocurrent pulse response has two characteristic regions: fast and slow [44]. The fast part starts immediately after illumination is switched on and associated with the process of establishment of thermal equilibrium between the shallow level and conduction band. The second (slow) region is observed under quasi steady-state condition and has characteristic time $\tau'(0)$.

The frequency dependence of the photo-EMF signal for $\omega > \beta$, as seen from Fig. 1.2, has another frequency independent region lying between two characteristic frequencies. The first one is equal to

$$\omega_3 \approx \beta \tag{1.43}$$

and the second one (for $\beta \gg \tau_M^{-1}$) can be written as:

$$\omega_4 \approx \frac{1 + \Theta + K^2 L_D^2}{\tau_D} + \frac{1}{\tau_T}. \tag{1.44}$$

One can see that the expressions for ω_2 and ω_4 have similar forms for the case of "relaxation type" photoconductor ($\tau'(0) \ll \tau_M$), linear photoconductivity ($\Theta \ll 1$) and low spatial frequencies ($KL_D < 1$). The second cut-off frequency ω_2 equals the inverse relaxation time of photoconductivity, i.e., $\omega_2 \approx \tau'(0)^{-1}$, whereas the fourth one ω_4 equals the inverse actual lifetime of electrons in the conduction band, i.e., $\omega_4 \approx \tau_D^{-1} + \tau_T^{-1}$.

The photocurrent amplitude in the frequency range $\omega_3 < \omega < \omega_4$ can be written as:

$$j_2 \approx \frac{-0.5m^2 \Delta\sigma_0 E_d (1 + \Theta + K^2 L_S^2)^{-1}}{1 + \Theta + \tau_D/\tau_M + \tau_D/\tau_T + K^2 L_D^2}. \tag{1.45}$$

The existence of such a plateau on the frequency transfer functions of the photo-EMF (Fig. 1.2) and photoconductivity [Fig. 1.4(a)] can be explained as follows: the analysis of the effect shows that photocurrent signal is proportional to $(\omega\tau')^{-1}$ for modulation frequencies $\omega > \omega_2$. In the frequency range $\beta < \omega < \tau_T^{-1}$ the value $|\tau'|$ depends on frequency as ω^{-1} [see Eq. (1.32)] so the photocurrent and photoconductivity amplitudes remain approximately constant here.

The frequency dependencies of complex amplitudes N and M are shown in Fig. 1.5. As seen dependencies $|M^{++}(\omega)|$ and $|n^{++}(\omega)|$ are similar in the frequency range $\omega < \omega_3$ (in fact, $M^{++} \approx n^{++} T_T/\tau_T$). It means that space charge waves $en^{++} \exp(iKx + i\omega t)$ and $eM^{++} \exp(iKx + i\omega t)$ are summed

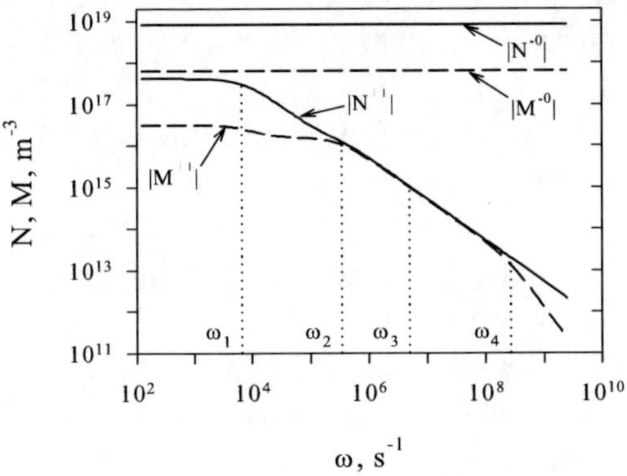

Figure 1.5. Amplitudes of densities N and M versus frequency of phase modulation ω ($I_0 = 10^4$ W/m^2, $K = 10^6$ m^{-1}, $m = 0.2$, $\Delta = 0.1$).

for these frequencies and this sum can be considered as an effective wave of space charge traveling in the conduction band of the crystal characterized with the effective parameters μ', γ'_D. For frequencies $\omega > \omega_4$ we have $n^{++} \approx N^{++}$.

For the frequency range $\omega \ll \beta$ we can use the results obtained before, since expression (1.30) with $\tau' = \tau'(0)$ has the form similar to Eq. (19) in Ref. [42]. There are still, however, minor differences associated with definition of L_S and Θ. The photo-EMF signal in this frequency range can be characterized by maximum amplitude j_1 and corresponding cut-off frequencies ω_1, ω_2. [42] At these frequencies the photocurrent amplitude decreases by a factor of $1/\sqrt{2}$ (Fig. 1.2) and its phase equals to $-3\pi/4$ and $3\pi/4$ (Fig. 1.3).

The maximum amplitude j_1 and the cut-off frequencies ω_1, ω_2 are the values of interest since having been measured they allow to estimate the Maxwell relaxation time τ_M, effective lifetime $\tau'(0)$, diffusion length L_D, parameter Θ and the effective mobility $\mu' = eL_D^2/[k_B T\tau'(0)]$. Parameters τ_M, $\tau'(0)$, L_D can be obtained from the spatial frequency dependencies of characteristic frequencies $\omega_1(K)$ and $\omega_2(K)$. For the case of "relaxation type" photoconductor ($\tau'(0) \ll \tau_M$), linear photoconductivity ($\Theta \ll 1$) and low spatial frequencies ($KL_D, KL_S < 1$) we have $\omega_1 \approx \tau_M^{-1}$, $\omega_2 \approx \tau'(0)^{-1}$, and at spatial frequency $K = L_D^{-1}$ these cut-off frequencies change by a factor of 2: $\omega_1(K = L_D^{-1}) = 0.5\omega_1(K \approx 0)$, $\omega_2(K = L_D^{-1}) = 2\omega_2(K \approx 0)$. The value of diffusion length L_D can be also directly estimated from the dependence $j_1(K)$. As it follows from Eqs. (1.15), (1.16), (1.25), (1.34), (1.41) the value Θ can be expressed as the following ratio $\Theta = \tau'(0)/T_D$. So, if the photoconductivity σ_0 depends on light intensity as $\sigma_0(I_0) \propto I_0^x$ (in the finite region near I_0) then $\Theta(I_0) = 1/x - 1$. This means that $\Theta(I_0)$ can be estimated from the intensity dependence $\omega_1(I_0)$ measured for low spatial frequency ($KL_D, KL_S < 1$) in the crystal with $\tau'(0) \ll \tau_M$. Intensity dependence $\omega_2(I_0)$ can provide information about Θ as well. Finally, for crystal with $\tau'(0) \ll \tau_M$ and low spatial frequency ($KL_D < 1$) the value Θ can be estimated as follows: $\Theta(I_0) = [\omega_2(I_0) - \omega_2(0)]/[\omega_2(I_0) + \omega_2(0)]$.

As seen from Figs. 1.2, 1.3 and Eq. (1.43) there is a possibility for the experimental measurement of the thermal excitation rate from the shallow level β: at frequency $\omega = \beta$ the photocurrent equals $|j^\omega| = \sqrt{2}|j_2|$ and its phase is $3\pi/4$. Furthermore, from the measured values of corresponding cut-off frequencies ω_3, ω_4 and using Eqs. (1.26), (1.41), (1.43), (1.44) one can easily estimate the value of the electron lifetime with respect to the deep donor level

$$\tau_D \approx [1 + \Theta + K^2L_D^2 + \omega_3\tau'(0)]/(\omega_3 + \omega_4) \tag{1.46}$$

and the lifetime of electrons in the conduction band

$$\tau = (\tau_D^{-1} + \tau_T^{-1})^{-1} = [\omega_4 - (\Theta + K^2L_D^2)/\tau_D]^{-1}. \tag{1.47}$$

Besides that, from the estimated values of the diffusion length L_D and lifetime τ_D, we can easily obtain the real (band) mobility of photocarriers: $\mu = eL_D^2/[k_BT\tau_D]$.

Let us analyze Equation (1.30) for "relaxation type" photoconductor ($\tau'(0) \ll \tau_M$), for the case of linear recombination ($\Theta \ll 1$) and for low frequency range $\omega \ll \beta$. The most interesting dependencies $j^\omega(K)$ are those obtained for $\omega \ll \omega_1$ and for $\omega = \omega_{m1}$ (i.e., $j_1(K)$).

In the frequency range $\omega \ll \omega_1$ the signal amplitude can be written as:

$$j^\omega(K) \propto K(1 + K^2L_S^2)^{-2}. \tag{1.48}$$

The signal amplitude grows linearly versus spatial frequency for $KL_S < 1/\sqrt{3}$ (Fig. 1.6, line c) which is due to increase of diffusion field amplitude E_d and decreases as $\sim K^{-3}$ for $KL_S > 1/\sqrt{3}$, behavior corresponding to saturation of deep traps. The maximum on this dependence is observed for $K = (\sqrt{3}L_S)^{-1}$.

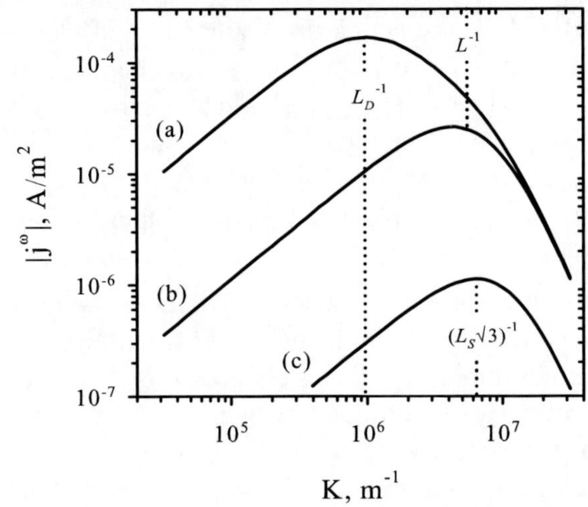

Figure 1.6. Dependence of the photocurrent amplitude j^{ω} on spatial frequency K calculated for characteristic temporal frequencies: (a) $\omega_1 < \omega < \omega_2$ [Eq. (1.39)], (b) $\omega_3 < \omega < \omega_4$ [Eq. (1.45)], (c) $\omega \ll \omega_1$ [Eq. (1.30)] for $\omega/2\pi = 2$ Hz; $I_0 = 10^4$ W/m^2, $m = 0.2$, $\Delta = 0.1$.

For $\omega = \omega_{m1}$ as seen from Eq. (1.39) we have the following expression for the signal amplitude

$$j^{\omega}(K) \propto K(1 + K^2 L_S^2)^{-1}(1 + K^2 L_D^2)^{-1}. \tag{1.49}$$

In photorefractive sillenite crystals the value of the diffusion length of electrons is usually larger than the Debye screening length, i.e., $L_D > L_S$. In this case the decrease of the signal amplitude as $\sim K^{-1}$ (see dependence $j_1(K)$ in Fig. 1.6, line a) is due to smoothing of electron density distribution in the conduction band. The photocurrent amplitude peaks at spatial frequency $K \approx L_D^{-1}$ [20].

Let us analyze the spatial frequency dependence $j_2(K)$ of the photo-EMF signal for high excitation frequencies $\omega_3 < \omega < \omega_4$. For $\Theta \ll 1$, $\tau_D \ll \tau'(0) \ll \tau_M$ Equation (1.45) can be written in the following form:

$$j^{\omega}(K) \propto K(1 + K^2 L_S^2)^{-1}(1 + K^2 L^2)^{-1}. \tag{1.50}$$

Here $L = \sqrt{\mu\tau k_B T/e}$ is the actual diffusion length of electrons. If the condition $L > L_S$ is satisfied the maximum on this dependence is observed at $K \approx L^{-1}$ (Fig. 1.6, line b). This means that real parameters can be determined from the spatial frequency dependence of the photo-EMF signal measured in the frequency range $\omega_3 < \omega < \omega_4$. Such practical estimation of real diffusion length for sillenites may be complicated because of approaching L to L_S (for the chosen material parameters (1.35) $L = 1.8 \times 10^{-7}$ m and $L_S = 9.1 \times 10^{-8}$ m).

Let us first consider the intensity dependencies of the photocurrent amplitude and corresponding cut-off frequencies for low illumination levels. The shallow level is almost empty in this case [see Eq. (1.21)] and all intensity dependencies of the effect are determined by behavior of the deep level. It follows from the fact that the intensity dependencies of values n_0, σ_0, τ_M, τ_D, $\tau'(0)$, L_D, L_S, Θ are expressed via the intensity dependence of the density N_0 [see Eqs. (1.16), (1.25), (1.31) - (1.34)]. The dependence $N_0(I_0)$ (1.24) in its turn is almost identical to the analogous dependence for the standard model of semiconductor with one deep level [42]. The difference is described by the presence of the intensity independent factor $(1 + M_T/N_{CM})$ and can be eliminated by using effective parameter γ'_D (1.27).

For the case of linear recombination of photocarriers ($N_0 \ll N_A$) and for $\tau'(0) \ll \tau_M$ corresponding values j_1, j_2 and ω_1 [Eqs. (1.39), (1.45), (1.37)] are proportional to the average light intensity I_0, other quantities, i.e., ω_2, ω_3, ω_4 [Eqs. (1.38), (1.43), (1.44)], are intensity independent. It follows from the linear character of the photoconductivity $\tau_M^{-1} \propto \sigma_0 \propto I_0$ and intensity independence of the other material parameters: $\tau'(0), \tau_D, \tau_T, L_D, L_S, \Theta \approx \text{const}(I_0)$.

The situation of quadratic recombination ($N_0 \gg N_A$) and for $\tau'(0) \sim \tau_M$ is more complicated: all values except ω_3 are intensity dependent. This peculiarity appears due to the intensity dependencies of the material parameters: σ_0, τ_M^{-1}, τ_D^{-1}, $\tau'(0)^{-1}$, L_D^{-2}, $L_S^{-2} \propto I_0^{0.5}$. For low spatial frequencies ($KL_D, KL_S < 1$) we have $j_1, \omega_1, \omega_2 \propto I_0^{0.5}$, the character of dependencies $j_2(I_0), \omega_4(I_0)$ can vary according to the relation between τ_D and τ_T. For the sillenite crystals sublinear photoconductivity dependence is observed [47], $\tau_T \ll \tau_D$ ($\tau_T = 4 \times 10^{-9}$ s [2], $\tau_D \sim (3 - 160) \times 10^{-6}$ s) and hence $j_2 \propto I_0$, $\omega_4 \approx \text{const}(I_0)$. The value τ_D is calculated using Eq. (1.41) and experimental estimations of the photoconductivity relaxation time $\tau'(0) = (0.15 - 8) \times 10^{-3}$ s obtained for the sillenite crystals grown in an oxygen-free atmosphere [46, 47].

1.3. Strong Filling of Shallow Levels

Let us consider the signal dependence for the case of sufficient filling of the shallow level. In order to perform numerical calculations of photovoltage we have to estimate stationary and spatially uniform densities N_0, M_0 and n_0. Let us assume that the conditions of linear generation (1.22) and linear recombination

$$N_0 \ll N_A \tag{1.51}$$

are satisfied for the deep level. Then the necessary concentrations are easily calculated from (1.1), (1.2), (1.20) and equal to

$$n_0 = S_D I_0 (N_D - N_A)/(\gamma_D N_A), \tag{1.52}$$

$$M_0 = \frac{M_T}{1 + (\beta + S_T I_0)/(\gamma_T n_0)}, \tag{1.53}$$

and the quantity N_0 is defined as their sum (1.20). As seen from Eq. (1.52) the concentration of electrons in the conduction band (average photoconductivity) is proportional to the light intensity while M_0 tends to the finite value for high light intensity (compare to Eq. (15) in Ref. [41]):

$$M_\infty = \lim_{I_0 \to \infty} M_0 = \frac{M_T}{1 + [S_T N_A \gamma_D]/[S_D (N_D - N_A)\gamma_T]}. \tag{1.54}$$

For the case

$$S_D(N_D - N_A)/S_T M_T \gg \gamma_D N_A / \gamma_T M_T \tag{1.55}$$

this value is close to the total density of shallow traps M_T, so the strong filling of shallow level is possible for high light intensities.

In this section we analyse the effect of photo-EMF for high light intensity when the shallow level is filled substantially:

$$M_0 \approx M_T. \tag{1.56}$$

For arbitrary material parameters and occupancy degrees of both levels one should use the general expression [Eq. (1.19)] to calculate the photocurrent amplitude. Here we shall consider two interesting situations when Eq. (1.19) reduces to the form similar to the one-level model approximation.

Let us assume that the light excitation from the deep level strongly exceeds the light excitation from the shallow one [Eqs. (1.28), (1.55)]. In this case the shallow level is practically totally filled with electrons. For this reason we can neglect in Eq. (1.19) all terms containing parameters of shallow traps and we obtain the standard expression for photocurrent amplitude:

$$j^{\omega} = \frac{-0.5m^2\Delta\sigma_0 E_d(1+\Theta+K^2L_S^2)^{-1}i\omega\tau_M}{1+\Theta+K^2L_S^2-\omega^2\tau_M\tau_D+i\omega[\tau_D+\tau_M(1+\Theta+K^2L_D^2)]}. \quad (1.57)$$

Here Θ is given by

$$\Theta = n_0/[(N_A+N_0)(1-N_A/N_D)]. \quad (1.58)$$

Equation (1.57) is identical to the photocurrent amplitude calculated for one-level model of semiconductor [42]. In this case, however, investigation of frequency transfer function $j^{\omega}(\omega)$ and dependence $j^{\omega}(K)$ can provide estimations of real (but not effective) values τ_D, L_D and μ.

We have estimated the minimal light intensity needed for sufficient filling of shallow level so that Eq. (1.57) is the consequence of common expression (1.19) in whole frequency range: $I_0 \sim 3\times10^6$ W/m^2 (this estimation was obtained using Eqs. (1.20), (1.52), (1.53) for stationary concentrations, material parameters (1.35) and assuming $M_T = 1\times10^{20}$ m^{-3}, $S_T = 0$). For this intensity the shallow level is filled by 95%.

Let us consider the situation when significant light generation of electrons from the shallow level takes place:

$$S_T M_0 \sim S_D(N_D - N_A). \quad (1.59)$$

If $\gamma_T M_T \gg \gamma_D N_A$ the shallow level is practically filled for high light intensities [see Eq. (1.54)] and we can neglect the items containing τ_T in Eq. (1.19). The expression for the complex amplitude of photo-EMF can be written then as follows:

$$j^{\omega} = \frac{-0.5m^2\Delta\sigma_0 E_d(1+\Theta+K^2L_S^2)^{-1}i\omega\tau_M A(0)A(\omega)}{1+\Theta+K^2L_S^2-\omega^2\tau_M\tau_D+i\omega[\tau_D+\tau_M(1+\Theta+K^2L_D^2)]}, \quad (1.60)$$

where

$$A(\omega) = 1 + \frac{S_T M_0 T_T}{S_D(N_D-N_A)T_D}\frac{1+i\omega T_D}{1+i\omega T_T}. \quad (1.61)$$

The frequency transfer function described by Eqs. (1.60), (1.61) is presented in Fig. 1.7 (curves b, c). The calculations were performed using Eqs. (1.20), (1.52), (1.53) for stationary concentrations and material parameters (1.35). We also assumed that $M_T = 1 \times 10^{20}$ m^{-3}, $S_T = 10$ m^{2}J^{-1} (choosing such S_T we satisfy condition (1.59)). The characteristic cut-off frequencies are given by the following expressions:

$$\omega_1 \approx \frac{1+\Theta+K^2L_S^2}{\tau_D+\tau_M(1+\Theta+K^2L_D^2)}, \tag{1.62}$$

$$\omega_2 \approx \frac{1}{T_T[1+S_TM_0/S_D(N_D-N_A)]}, \tag{1.63}$$

$$\omega_3 \approx \frac{1}{T_T+\tau_D/(1+\Theta+\tau_D/\tau_M+K^2L_D^2)}, \tag{1.64}$$

$$\omega_4 \approx \frac{1}{\tau_M}+\frac{1+\Theta+K^2L_D^2}{\tau_D}+\frac{1}{T_T} + \frac{1}{T_T+\tau_D/(1+\Theta+\tau_D/\tau_M+K^2L_D^2)}. \tag{1.65}$$

Calculating ω_2, ω_3, ω_4 we have supposed $T_T \ll \tau_D + \tau_M(1+\Theta+K^2L_D^2)$. For the considered range of light intensities $I_0 = 0 - 10^8$ W/m^2 the following relation is satisfied: $\tau_M \geq 30T_T$. The amplitude of the photo-EMF signal for the frequency range $\omega_3 < \omega < \omega_4$ can be written as:

$$j_2 \approx \frac{-0.5m^2\Delta\sigma_0E_dA(0)[1+S_TM_0/S_D(N_D-N_A)]}{(1+\Theta+K^2L_S^2)(1+\Theta+\tau_D/\tau_M+\tau_D/T_T+K^2L_D^2)}. \tag{1.66}$$

For low frequencies of phase modulation ($\omega < \omega_2$) the effect of photo-EMF is described by standard expression (1.57) since $A(0) \approx 1$. As seen from Fig. 1.7 the presence of shallow traps reveals itself only for high excitation frequencies ($\omega > \omega_2$). In contrast to the low intensity approximation ($M_0 \ll M_T$, $S_TM_0 \ll S_D(N_D-N_A)$) the influence of the shallow level for the case of high light intensities ($M_0 \approx M_T$, $S_TM_0 \sim S_D(N_D-N_A)$) leads to the increase of photocurrent at high frequencies of phase modulation (compare Figs. 1.2 and 1.7).

This fact can be explained as follows: when the shallow level is filled slightly it play a role of an electron reservoir leading to the increase of the

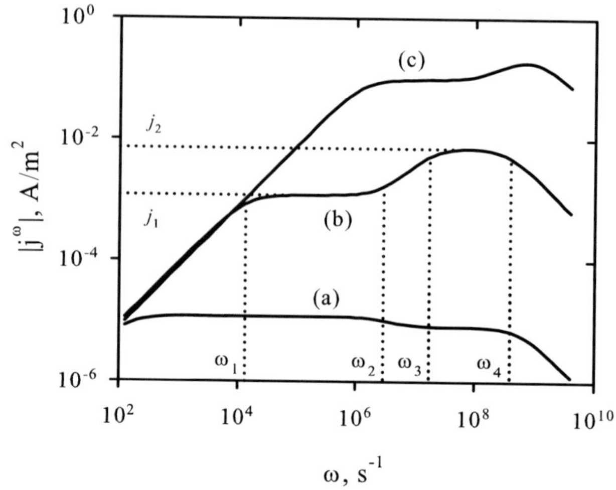

Figure 1.7. Frequency transfer function of the photo-EMF for different light intensities $I_0 = 3 \times 10^3$, 3×10^5, 3×10^7 W/m^2 (curves (a), (b) and (c) respectively). The calculations are performed using Eq. (1.19) for $K = 6.3 \times 10^6$ m^{-1}, $m = 0.2$, $\Delta = 0.1$.

photoconductivity relaxation time and reducing the effective electron mobility. However, when this level is filled strongly the characteristic time τ_D tends to its actual value [see Eqs. (1.57), (1.60)] and sufficient light generation from this level leads to photocurrent increase with respect to standard estimation of (1.57).

In the first part of the Chapter we have analyzed the influence of shallow traps on the effect of non-steady-state photo-EMF. We have derived the general expression for amplitude of the photocurrent for apparent filling degrees of deep and shallow levels. We analyze the photocurrent behavior with respect to temporal and spatial frequencies and discuss the case of high illumination levels. The general calculations were performed for the crystal's parameters typical for photorefractive $Bi_{12}SiO_{20}$. The influence of shallow traps is not restricted to considered cases. The effect of non-steady-state photo-EMF in other photoconductive crystals (e.g., semi-insulating GaAs) with different material constants, may reveal peculiarities at other characteristic temporal and spatial frequencies.

2. Drift Regime

The subject of space charge waves covers a wide range of physics from plasmas, acoustics and semiconductors to the relatively new area of photorefractive optics [48, 49]. Space charge waves can be regarded as quasiparticles in solids and are of great interest not only in the case of photorefractive materials but also in the case of many semiconductors. They determine the dynamic behavior of the space charge distribution in the crystal and can be excited optically by various methods [45, 50–52]. A powerful technique is to use the spatially oscillating interference pattern created by two coherent laser beams. Then, if the frequency of oscillation and the grating spacing of the interference pattern coincide with the frequency and the wavelength of the corresponding space charge wave, a resonance excitation of the space charge wave occurs. Space charge and photoconductivity waves play important role in inducing non-steady-state photocurrents in photorefractive crystals [20, 38]. This effect provide an unique opportunity for direct measurement of the photoelectron's drift mobility from the position of the corresponding resonance maxima in frequency transfer function of the photocurrent [38, 53].

In this part of the Chapter we shall analyze the resonant excitation of space charge and photoconductivity waves and their detection using the non-steady-state photoelectromotive force effect. We consider the model of semiconductor with shallow energy level and present numerical calculations for the commonly used $Bi_{12}SiO_{20}$ crystal. It is important to point out that application of the external electric field completely changes the physics of the phenomenon and here we deal with the *resonant* excitation of the *eigen-waves* in the material and not with the *forced* excitation of the effect as in Ref. [20].

Let us consider the model of semiconductor with two levels from which photoelectrons are generated [Fig. 1.8]. We confine ourselves to the photoexcitation from the deep donor level (with the total density N_D) and both light and thermal excitation from the shallow level (with the total density M_T). The shallow level is assumed to be empty in the dark and the deep one is partially filled (with the density of vacant states N_A).

Suppose a crystal is illuminated with the oscillating interference pattern $I(x,t) = I_0\{1 + m\cos[Kx + \Delta\cos(\omega t)]\}$. Theoretical analysis of the non-steady-state photocurrent generation is performed for the following simplify-

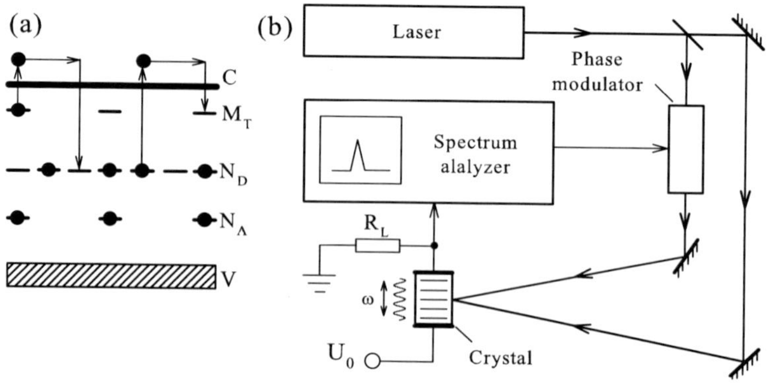

Figure 1.8. Model of semiconductor with shallow levels (a) and experimental setup for measurements of the non-steady-state photocurrent (b).

ing assumptions: a negligible dark conductivity, a low contrast light pattern ($m \ll 1$), and low-amplitude vibrations of the interference pattern ($\Delta \ll 1$). No assumptions were made on the occupancy of both levels and the prevalence of any activation mechanism for the shallow level.

We have obtained the general expression for the photocurrent amplitude. It is too bulky, however, so we shall discuss the most interesting case of the linear generation from the deep level, low filling degree and negligible light excitation from the shallow level:

$$N_0 \ll N_D - N_A, \quad M_0 \ll M_T, \quad S_T I_0 \ll \beta,$$
$$S_T M_0 \ll S_D(N_D - N_A), \quad \beta T_D \gg 1. \tag{1.67}$$

Here N_0, M_0, n_0 are the stationary and spatially uniform densities (defined from Eqs. (1.1), (1.2) and neutrality condition $N_0 = M_0 + n_0$), $T_D = (S_D I_0 + \gamma_D n_0)^{-1} \approx (\gamma_D n_0)^{-1}$ is the inverse of the sum of the generation and ion recombination rates for the deep level. The fifth condition ($\beta T_D \gg 1$) states that the number of free carriers excited from shallow level during lifetime of ionized deep level is larger than the number of carriers captured at this level.

Under these assumptions the expression for the photocurrent amplitude is simplified to:

$$
\begin{aligned}
j^\omega &= \frac{0.25m^2\Delta\sigma_0(Z^*)^{-1}[i2E_0 - \omega\tau_M(E_0 + iE_d)]}{Z - \omega^2\tau'\tau_M + i\omega[\tau' + \tau_M(1 + \Theta + K^2L_D^2 + iKL_0)]} \\
&- \frac{0.25m^2\Delta\sigma_0 Z^{-1}[i2E_0 - \omega\tau_M(E_0 - iE_d)]}{Z^* - \omega^2\tau'\tau_M + i\omega[\tau' + \tau_M(1 + \Theta + K^2L_D^2 - iKL_0)]}. \quad (1.68)
\end{aligned}
$$

Here $\sigma_0 = e\mu n_0$ is the average photoconductivity of the crystal, E_0 is the external electric field, $E_d = Kk_BT/e$ is the diffusion field, $\tau_M = \epsilon\epsilon_0/\sigma_0$ is the Maxwell relaxation time,

$$
\tau' = \tau'(\omega) = \tau_D\left(1 + \frac{\tau_T^{-1}\beta^{-1}}{1 + i\omega\beta^{-1}}\right) \quad (1.69)
$$

is the complex and frequency-dependent effective electron lifetime, $\tau_D = [\gamma_D(N_A+N_0)]^{-1}$, $\tau_T = [\gamma_T(M_T-M_0)]^{-1}$ are the recombination times of electrons to the deep and shallow traps, $L_D = \sqrt{D\tau_D}$, $L_0 = \mu\tau_D E_0$ are the average diffusion and drift lengths of photocarriers, $L_S = [\epsilon\epsilon_0 k_BT/e^2(N_A + N_0)(1 - N_A/N_D)]^{1/2}$ is the Debye screening length [41], $Z = 1 + \Theta + K^2L_S^2(1 + iE_0/E_d)$ and $\Theta = [(1 + N_A/N_0)(1 - N_A/N_D)]^{-1}$ are the dimensionless parameters ($0 < \Theta < 1$).

The frequency dependence of the photovoltage amplitude is shown in Fig. 1.9. The material parameters used for numerical calculations are taken from Refs. [33, 45]. Depending on the amplitude of the applied electric field E_0, spatial frequency K and material parameters the frequency response of the photocurrent can have from one to three maxima.

Let us first consider the low-frequency range of photocurrent excitation when $\omega \ll \beta$. For such modulation frequencies the effective electron lifetime is the real value: $\tau' \approx \tau'(0) = \tau_D(1 + \tau_T^{-1}\beta^{-1})$. It can be shown that for $KL_0 > 1 + \Theta + K^2L_D^2$ there is a maximum on the frequency dependence of the non-steady-state photocurrent at $\omega_{r1} \approx |Z|/|\tau'(0) + \tau_M(1 + \Theta + K^2L_D^2 + iKL_0)|$. The existence of such a peak is associated with the excitation of space charge waves [50, 51]. For $L_S < L_D$ the resonant frequency ω_{r1} is a decreasing function of the applied electric field E_0 [38,50]. However, the finite screening length L_S taken into account makes ω_{r1} tend to a non-zero limit for high electric fields $E_0 \to \infty$: $\omega_{r1} \to T_D^{-1}$ (compare to Eq. (16) in Ref. [38]).

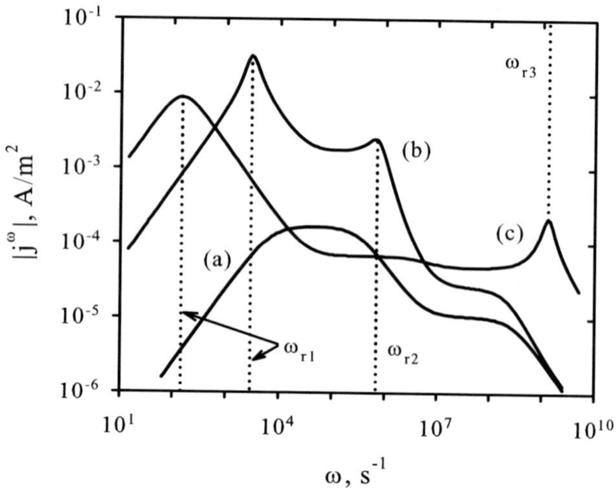

Figure 1.9. Theoretical dependencies of the photocurrent amplitude $|j^\omega|$ on the oscillation frequency ω ($I_0 = 10^4$ W/m^2, $m = 0.2$, $\Delta = 0.1$): (a) $E_0 = 0$, $K = 1.00 \times 10^6$ m^{-1}, (b) $E_0 = 6 \times 10^5$ V/m, $K = 1.89 \times 10^5$ m^{-1}, (c) $E_0 = 2 \times 10^6$ V/m, $K = 1.89 \times 10^6$ m^{-1}.

For the intermediate range of excitation frequencies when $\omega_{r1} \ll \omega \ll \beta$ and for the case of relaxation type semiconductor $[\tau'(0) \ll |\tau_M(1 + \Theta + K^2L_D^2 + iKL_0)|]$ the second maximum is observed on the frequency transfer function. Two types of this maximum can be considered in this case. The first one corresponds to the resonant excitation of a photoconductivity wave and is detected at $\omega_{r2} \approx KL_0/\tau'(0) = \mu' E_0 K$, where $\mu' = \mu/(1 + \tau_T^{-1}\beta^{-1})$ is the effective mobility of electrons. Such a resonant maximum exists in the frequency range $(1+\Theta+K^2L_D^2)/\tau'(0) < \omega < \beta$. This characteristic frequency ω_{r2} is an increasing function of values E_0 and K, but when it approaches the thermal excitation rate β, the corresponding peak disappears. Thus there is an optimum value of the external electric field that maximizes the amplitude of the resonant peak:

$$E_0 \approx \frac{\beta}{\mu' K}\sqrt{\frac{\tau_T}{\tau_D}(1 + \Theta + K^2L_D^2)}, \tag{1.70}$$

$$j^{\omega} \approx \frac{-im^2\Delta\sigma_0\beta}{8Z\mu' K}\sqrt{\frac{\tau_T}{\tau_D(1+\Theta+K^2L_D^2)}}. \tag{1.71}$$

The second type of the maximum can be observed at another characteristic frequency: $\omega_{m2} \approx (1+\Theta+K^2L_D^2)/\tau'(0)$. This peak on the frequency response of the photocurrent is detected only for specific relations between the amplitude of the applied electric field E_0 and spatial frequency of the interference pattern K: $KL_0, KL_D, KL_S \leq 1$, $L_0 \approx [(1+\Theta)(L_D^2+L_S^2)]^{0.5}$. It is important to point out that the origin of this maximum is not associated with the resonant excitation of photoconductivity waves. The characteristic frequency ω_{m2} weakly depends on spatial frequency K, and the amplitude of this maximum increases with increasing K value.

For high excitation frequencies ($\omega > \tau_T^{-1}$) the effective lifetime of electrons is the real value as well: $\tau' \approx \tau_D$. The third resonant peak is observed in this case at the following frequency:

$$\omega_{r3} \approx KL_0/\tau_D = \mu E_0 K. \tag{1.72}$$

The presence of such a peak is the characteristic feature of the discussed model. This maximum can be observed in following frequency region $\omega_{r3} > (1+\Theta+K^2L_D^2)\tau_D^{-1} + \tau_T^{-1}$. In other words, this condition means that the photoconductivity waves are resonantly excited if an electron in conduction band is neither captured by deep nor by shallow traps during the period of interference pattern oscillation. The values of effective μ' and actual μ mobilities can be easily obtained from characteristic resonant frequencies ω_{r2} and ω_{r3} on the frequency dependence of the non-steady-state photocurrent.

The non-steady-state photocurrent amplitude for the second and third resonances can be estimated from the following expression

$$j^{\omega}_{r2,r3} \approx \frac{-i0.25m^2\Delta\sigma_0 E_0 Z^{-1}}{1+\Theta+K^2L_D^2+\tau_D/\tau_T(1+\beta^2/\omega_{r2,r3}^2)}. \tag{1.73}$$

We have assumed here that $\omega_{r2}, \omega_{r3} \gg \omega_{r1}$, $|\tau'| \ll |\tau_M(1+\Theta+K^2L_D^2 + iKL_0)|$, $E_0 \gg E_d$.

The origin of the resonance maxima on the frequency transfer function can be explained as follows. The light intensity distribution $I(x,t)$, which is external periodic driven force in our case, contains the waves running in the opposite direction with respect to applied electric field ($E_0 > 0$): $I^{\pm\pm}\exp[\pm i(Kx+\omega t)]$. These light waves give rise the analogous waves of free carriers concentration: $n^{\pm\pm}\exp[\pm i(Kx+\omega t)]$. The phase velocities of these waves equal $v_{ph} = -\omega/K$. For $E_0 \gg E_d$ and $m \ll 1$ each separate electron migrates in semiconductor crystal with the average drift velocity $v_{dr} = -\mu E_0$ [during the actual electron lifetime $(\tau_D^{-1}+\tau_T^{-1})^{-1}$] and $v_{dr} = -\mu' E_0$ [during the effective electron lifetime $\tau'(0)$]. If the phase and drift velocities are not matched, i.e., $v_{ph} \neq v_{dr}$, this mentally selected electron outruns or lags behind the packet which phase velocity v_{ph} is determined by external influence $I(x,t)$. As the result, the concentration density distribution $n^{\pm\pm}\exp[\pm i(Kx+\omega t)]$ become blurred. At resonant frequencies ω_{r2}, ω_{r3} the photoconductivity wave and all the photocarriers migrate synchronously ($v_{ph} = v_{dr}$) giving rise to the appearance of corresponding resonant peaks. For the waves traveling in the direction of applied electric field, i.e., for $n^{\pm\mp}\exp[\pm i(Kx-\omega t)]$, the phase and drift velocities have the opposite signs. This means, in particular, that "smoothing" of the photoconductivity wave takes place for any modulation frequency ω and this component does not contribute to the resonant excitation of the external photocurrent.

Let us write down corresponding complex amplitude of the space-charge electric field formed in the crystal bulk for $E_0 \gg E_d$, $\Theta \ll 1$:

$$E^{++} = \frac{-i0.25m\Delta E_0}{Z^* - \omega^2\tau_M\tau' + i\omega[\tau' + \tau_M(1 - iKL_0)]}. \tag{1.74}$$

It is seen from Eq. (1.74) that the time-dependent electric field grating amplitude is strongly decreased for high excitation frequencies $\omega \gg \omega_{r1}$. For example, this amplitude calculated at ω_{r3} and for the parameters used in Fig. 9, line (c) is only 3×10^{-3} V/m. This fact may essentially complicate the observation of the discussed effects using conventional holographic techniques.

The conventional experimental setup for observation of the non-steady-state photocurrents is presented in Fig. 1.8(b). A standard single-mode He-Ne laser with an average power of $P_0 = 40$ mW ($\lambda = 633$ nm) was used as the basic source of coherent radiation for formation of a recording interference pattern. To form the interference pattern with a specified fringe spacing and to cause its sinusoidal vibrations within the sample volume, we used a conventional Twyman-Green interferometer. An electro-optic phase modulator ML-102A produced phase modulation of the laser beam. The non-steady-state pho-

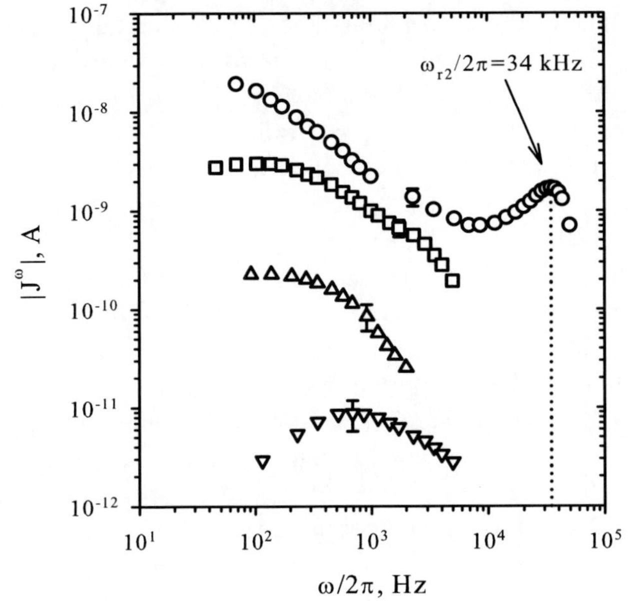

Figure 1.10. Frequency dependencies of photocurrent measured in $Bi_{12}SiO_{20}$ grown in argon atmosphere ($\lambda = 633$ nm, $K = 5.0 \times 10^4$ m^{-1}, $I_0 = 60$ W/m^2, $m = 0.2$, $\Delta = 0.2$): $E_0 = 0$ (∇), 2.0×10^5 ($\triangle$), 4.0×10^5 ($\square$), 1.0×10^6 V/m ($\circ$).

tocurrent was measured using standard SK4-56 spectrum analyzer. Figure 1.10 presents the frequency dependence of the non-steady-state photocurrent flowing in n-type photorefractive $Bi_{12}SiO_{20}$ grown in argon atmosphere. These crystals have several distinct features with respect to sillenite crystals grown in an air atmosphere [47]. First, these crystals have rather high photoconductivity in the red region of spectrum comparable with the photoconductivity of the usual sillenite crystals in the blue-green region of the spectrum [20]. Second, in all investigated crystals for moderate light intensity of the red light we observed unusual relation between the lifetime of photocarriers τ' and Maxwell relaxation time $\tau_M < \tau'(0)$. This means, in particular, that photocurrent measurements are done in the lifetime regime without external electric field. Nevertheless, the appropriate choice of the external electric field, makes the observation of the second resonance possible and photocurrent measurements are done in relaxation time regime $[\tau'(0) < \tau_M|1 + \Theta + K^2L_D^2 + iKL_0|]$. The drift mobility value was found to be $\mu' = (4.3 \pm 0.4) \times 10^{-6}$ m^2/V s ($I_0 = 60$ W/m^2) and corresponds to the values of effective mobilities observed in conventional sillenites [38]. Thus we think that the crystal growth in oxygen free atmosphere affect the deep levels mainly, changing the properties of shallow levels slightly. The mobility turned out to be dependent on the light intensity which can be associated with both filling of shallow levels and heating of the crystal in the external electric field.

In conclusion, we have reported resonant excitation of space-charge and photoconductivity waves in semiconductors with shallow energy levels. We predicted the appearance of several resonance maxima on the frequency transfer function of the photocurrent and showed that observation of photoconductivity wave with conventional holographic technique is unlikely.

Chapter 2

Influence of AC Electric Fields on the Non-Steady-State Photo-EMF

1. Nonresonant Case

Let us present investigations of the non-steady-state photocurrent generation in an external AC field. We show both experimentally and theoretically that the frequency transfer function of the effect is similar to the response of the effect in the diffusion regime of excitation, i.e., linear growth of the signal for low modulation frequencies up to a certain characteristic cut-off frequency and frequency-independent region for high modulation frequencies. The new technique for the signal enhancement and determination of the $\mu\tau$-product is proposed.

This section presents calculations of the non-steady-state photocurrent amplitude excited in an alternating electric field. Suppose a crystal is illuminated with the oscillating interference pattern formed by two plane light waves one of which is phase modulated with frequency ω and amplitude Δ. We consider the widely used model of semiconductor with one type of partially compensated donor levels [2]. The balance equation for the concentration of electrons in the conduction band n can be written as follows:

$$\frac{\partial n}{\partial t} = g - \frac{n}{\tau} + \frac{1}{e}\frac{\partial j_e}{\partial x}. \tag{2.1}$$

Here $g(x,t) = (\alpha\beta/h\nu)I(x,t)$ is the electron generation rate (α is the light absorption coefficient, β is the quantum efficiency, and $h\nu$ is the photon energy). We assume low illumination levels when the generation and recombination rates are linear functions of the light intensity and electron concentration respectively, i.e., for the considered model conditions $N_A + N \ll N_D$, $N_A + N \gg n$ are fulfilled [N_D is the total concentration of donors, N is the concentration of donor centers emptied by light, N_A is the concentration of the compensating acceptor states. In this case the electron lifetime is independent on the light intensity and determined by the concentration N_A only. The effects of trap saturation are neglected as well (screening lengths are small compared to the inverse value of the spatial frequency). In order to calculate the distributions of the electron concentration, electric field and resulting photocurrent Eq. (2.1) should be added by the continuity and Poisson equations and by the expressions for the charge ρ and electron current j_e densities:

$$\frac{\partial \rho}{\partial x} + \frac{\partial j_e}{\partial x} = 0, \tag{2.2}$$

$$\epsilon\epsilon_0 \frac{\partial E}{\partial x} = \rho, \tag{2.3}$$

$$\rho = e(N - n), \tag{2.4}$$

$$j_e = e\mu n E + k_B T \mu \frac{\partial n}{\partial x}. \tag{2.5}$$

where E is the electric field, μ is the electron mobility, ϵ is the dielectric constant, ϵ_0 is the free space permittivity, k_B is the Boltzmann constant, T is the temperature. Equations (2.1)-(2.5) can be represented as follows:

$$\frac{\partial n}{\partial t} = g - \frac{n}{\tau} - \frac{\epsilon\epsilon_0}{e}\frac{\partial^2 E}{\partial x \partial t}, \tag{2.6}$$

$$\frac{\partial}{\partial x}\left(e\mu n E + k_B T \mu \frac{\partial n}{\partial x} + \epsilon\epsilon_0 \frac{\partial E}{\partial t}\right) = 0. \tag{2.7}$$

The sinusoidal electric field with amplitude E_{ext} and frequency Ω is applied to the crystal: $E_{ext}(t) = E_{ext}\cos\Omega t$.

The calculation procedure simplifies substantially if we suppose the contrast m and amplitude of phase modulation Δ to be small: $m, \Delta \ll 1$. Then we can look for the solution of equations (2.6)-(2.7) to the lowest order with respect to the small parameters presenting $n(x,t)$, $E(x,t)$ as the sums of stationary and running gratings:

$$\begin{aligned} n &= n_0 + \sum_{p,q=-1}^{1} n^{+pq} \exp\{i[Kx + (p\omega + q\Omega)t]\} \\ &+ \sum_{p,q=-1}^{1} n^{-pq} \exp\{i[-Kx + (p\omega + q\Omega)t]\}, \end{aligned} \tag{2.8}$$

$$\begin{aligned} E &= E^{00+} \exp(i\Omega t) + E^{00-} \exp(-i\Omega t) \\ &+ \sum_{p,q=-1}^{1} E^{+pq} \exp\{i[Kx + (p\omega + q\Omega)t]\} \\ &+ \sum_{p,q=-1}^{1} E^{-pq} \exp\{i[-Kx + (p\omega + q\Omega)t]\}. \end{aligned} \tag{2.9}$$

It is important to point out that for the solution of similar tasks two basic approaches can be used: the utilization of truncated Fourier series [54] (as was also done in this paper) and the time averaging technique [55]. The former covers all time scales [56], the latter is convenient for arbitrary waveforms of the external field [57].

For the case of cyclic boundary conditions [20] the total current averaged over the interelectrode spacing L contains the drift and displacement components. The photocurrent signal is defined by the drift one: [20]

$$j(t) = \frac{1}{L}\int_0^L e\mu n(x,t)E(x,t)dx. \tag{2.10}$$

We confine ourselves to the analysis of the first harmonic of photocurrent (with frequency ω) which can be written in the form of the following combination of coefficients introduced in Eqs. (2.8), (2.9):

$$\begin{aligned} j^{\omega} &= 2e\mu \sum_{q=-1}^{1} \Big(n^{+1q} E^{-0(-q)} + n^{+0q} E^{-1(-q)} \\ &+ n^{-1q} E^{+0(-q)} + n^{-0q} E^{+1(-q)} \Big). \end{aligned} \tag{2.11}$$

The higher spatial and temporal harmonics (and even subharmonics) of values $n(x,t)$, $E(x,t)$ and resulting photocurrent $j(t)$ become substantial only for the case of large contrast and amplitude of phase modulation: $m, \Delta \sim 1$. [2, 5, 58] Although these problems are rather interesting and widely discussed in the literature the most valuable applications of the holographic photocurrents for the semiconductor material characterization are based on the measurements of the first harmonic of the signal.

Let us consider the case of "relaxation type" semiconductor , i.e., $\tau \ll \tau_M$ [$\tau_M = \epsilon\epsilon_0/e\mu n_0$ is the Maxwell relaxation time]. We suppose that the frequency of phase modulation ω is much smaller than the frequency of the external electric field Ω; the period of the AC field is larger than the carrier lifetime τ and shorter than grating relaxation time τ_M:

$$\omega \ll \Omega, \tag{2.12}$$

$$\tau_M^{-1} \ll \Omega \ll \tau^{-1}. \tag{2.13}$$

We have obtained all complex amplitudes of the electron concentration and electric field necessary for calculation of the first harmonic of the photocurrent signal. Some of them are given below:

$$\begin{aligned}
n^{+00} &= \frac{m}{2}n_0, \quad n^{+01} = \frac{imKL_0}{2\sqrt{2}(1+K^2L_D^2)}n_0,\\
n^{+10} &= \frac{im\Delta(1+i\omega\tau_M)(1+K^2L_D^2)n_0/4}{1+K^2L_D^2+i\omega\tau_M[(1+K^2L_D^2)^2+K^2L_0^2]},\\
n^{+11} &= \frac{-m\Delta(1+i\omega\tau_M)KL_0n_0/4\sqrt{2}}{1+K^2L_D^2+i\omega\tau_M[(1+K^2L_D^2)^2+K^2L_0^2]},\\
E^{+00} &= -i\frac{m}{2}\left(E_D+E_0\frac{KL_0}{1+K^2L_D^2}\right),\\
E^{+01} &= \frac{imE_0}{2\sqrt{2}\Omega\tau_M(1+K^2L_D^2)},\\
E^{+10} &= \frac{m\Delta[E_D(1+K^2L_D^2)+E_0KL_0]/4}{1+K^2L_D^2+i\omega\tau_M[(1+K^2L_D^2)^2+K^2L_0^2]},\\
E^{+11} &= \frac{-m\Delta(1+i\omega\tau_M)E_0/4\sqrt{2}\Omega\tau_M}{1+K^2L_D^2+i\omega\tau_M[(1+K^2L_D^2)^2+K^2L_0^2]}.
\end{aligned} \tag{2.14}$$

Here $E_0 = E_{ext}/\sqrt{2}$ is the effective value of the external electric field, $E_D = (k_B T/e)K$ is the diffusion field, $\sigma_0 = e\mu n_0$ is the average conductivity, $L_0 = \mu\tau E_0$ is the drift length of photocarriers, $L_D = (k_B T\mu\tau/e)^{1/2}$ is the diffusion length.

Finally the expression for the complex amplitude of the photocurrent can be written as:

$$j^\omega = \frac{-0.5m^2\Delta\sigma_0[E_D + E_0 K L_0/(1+K^2L_D^2)]i\omega\tau_M}{1+i\omega\tau_M[1+K^2L_D^2+K^2L_0^2/(1+K^2L_D^2)]}. \tag{2.15}$$

Let us list the main features of the nonstationary holographic photocurrents excited in the AC external field. As follows from Eq. (2.15) the frequency transfer function of the effect has the traditional form typical for the diffusion mechanism of photocurrent excitation, namely, the linear growth of the signal for low modulation frequencies and the frequency-independent region for high frequencies of phase modulation [20]. These two regions are separated by corresponding cut-off frequency:

$$\omega_0 = \frac{1}{\tau_M[1+K^2L_D^2+K^2L_0^2/(1+K^2L_D^2)]}. \tag{2.16}$$

For low spatial frequencies of the interference pattern (i.e., for $K^2L_D^2 \ll 1$) the most striking peculiarities can be observed. In this case the expressions for the photocurrent amplitude and corresponding cut-off frequency can be written as follows:

$$\begin{aligned} j^\omega &= \frac{m^2\Delta}{2}\sigma_0 E_D \frac{-i\omega\tau_M(1+L_0^2/L_D^2)}{1+i\omega\tau_M(1+K^2L_0^2)} \\ &= \frac{m^2\Delta}{2}\sigma_0 E_D \frac{-i\omega\tau_M(1+E_0^2/E_L^2)}{1+i\omega\tau_M(1+E_0^2/E_M^2)}, \end{aligned} \tag{2.17}$$

$$\begin{aligned} \omega_0 &= [\tau_M(1+K^2L_0^2)]^{-1} \\ &= [\tau_M(1+E_0^2/E_M^2)]^{-1}. \end{aligned} \tag{2.18}$$

Here $E_L = k_B T/eL_D$, $E_M = (K\mu\tau)^{-1}$ are the characteristic values of the electric field [45]. As seen from Eq. (2.17) the dependence of the photocurrent amplitude on electric field has a S-like shape. The signal grows as E_0^2 between

E_L and E_M due to the corresponding increase of the electric field grating amplitudes E^{+00}, E^{+10} [Eq. (2.14)]. Beyond this region ($E_0 < E_L$, $E_0 > E_M$) the signal is independent on the electric field. The external field does not affect the photocurrent amplitude while the effective drift length is smaller than the diffusion one ($L_0 \leq L_D$). For high values of the applied electric field (the so-called long drift length regime) an enhanced transfer of electrons characterized by the large effective drift length L_0 takes place. Under these conditions the nonstationary photoconductivity grating is blurred leading to the saturation of the signal amplitude. The corresponding cut-off frequency ω_0 [Eq. (2.18)] falls down when drift length L_0 becomes comparable to the spatial period of the recorded grating ($L_0 \geq K^{-1}$ or $E_0 \geq E_M$).

The dependencies of the photocurrent amplitude j^ω and cut-off frequency ω_0 versus the AC field amplitude E_0 can be used for the determination of transport parameters of photoexcited carriers in semiconductor materials. Indeed, the characteristic breakpoints in these dependencies give the values of the diffusion and drift lengths of photocarriers and consequently – the $\mu\tau$-product value.

We have also calculated the maximum gain of the photocurrent signal $j^\omega(\omega > \omega_0)$ which can be obtained in the sinusoidal AC field:

$$G_m = \frac{\lim\limits_{E_0 \to \infty} j^\omega}{\lim\limits_{E_0 \to 0} j^\omega} = \frac{1 + K^2 L_D^2 + L_0^2/L_D^2}{1 + K^2 L_D^2 + K^2 L_0^2/(1 + K^2 L_D^2)}. \tag{2.19}$$

It depends on the spatial frequency of the interference pattern: $G_m = 1/K^2 L_D^2$ for $K^2 L_D^2 \ll 1$ and $G_m = 1$ for $K^2 L_D^2 \gg 1$.

The experimental setup used for the measurements of the nonstationary holographic currents under applied alternating electric field is shown in Fig. 2.1. A standard He-Cd laser with an average power of $P_{out} \approx 1$ mW ($\lambda = 442$ nm) was used as the basic source of coherent radiation for formation of a recording interference pattern. To form the interference pattern with a specified fringe spacing and to cause its sinusoidal vibrations within the sample volume, we used a conventional Twyman-Green interferometer. An electrooptic phase modulator ML-102A produced phase modulation of the laser beam (with frequency ω and amplitude $\Delta = 0.16$). The voltage from generator was amplified by conventional transformer and then applied to the crystal. The photocurrent signal was filtered and then measured by spectrum analyzer SK4-56 ($f = 0.01 - 50$

kHz, $\Delta f = 3$ Hz). In our experiments we used n-type $Bi_{12}SiO_{20}$ crystal with characteristic dimensions $1 \times 3 \times 10$ mm^3, the front and back surfaces of the crystal (1×10 mm^2) were polished, the silver paste electrodes (3×4 mm^2) were painted on the lateral surfaces. The interelectrode spacing was 1 mm. In order to change the parameters of the material (concentration of the trapping centers, characteristic recombination times, etc. [44]) the crystal was simultaneously illuminated by the light from the microscope lamp focused by lens and passed through the infrared filter KS-15 ($\lambda_{IR} = 650 - 2700$ nm, $I_{IR} = 13$ mW/mm^2).

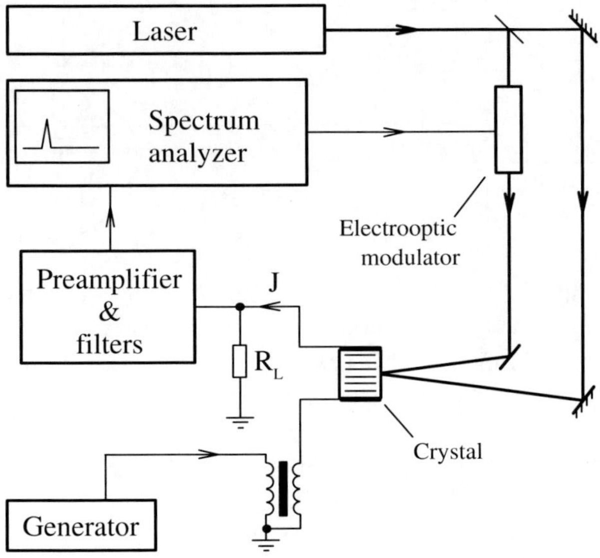

Figure 2.1. Experimental setup for measurements of the nonstationary holographic photocurrents in the presence of alternating electric field.

Let us present the results of the photocurrent measurements carried out in the external electric field and compare them with the data obtained for the diffusion excitation regime. Fig. 2.2 presents the frequency transfer function of the holographic photocurrent excited in $Bi_{12}SiO_{20}$ crystal in the absence of an external electric field. Recall that the frequency response of the effect is usually characterized by the corresponding cut-off frequencies [20]:

$$\omega_0 = [\tau_M(1+K^2L_D^2)]^{-1}, \quad (2.20)$$

$$\omega_0' = (1+K^2L_D^2)\tau^{-1}. \quad (2.21)$$

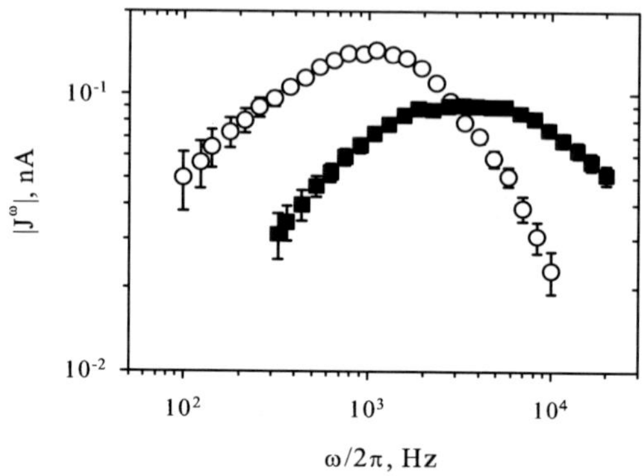

Figure 2.2. Frequency transfer functions of the nonstationary photocurrent measured with (squares) and without (circles) additional infrared illumination ($Bi_{12}SiO_{20}$, $\lambda = 442$ nm, $E_0 = 0$, $P_0 = 0.32$ mW, $m = 0.65$, $K = 5 \times 10^4$ m^{-1}).

From the dependencies of the photocurrent amplitude measured at low spatial frequency of the interference pattern ($K^2L_D^2 \ll 1$) and without additional illumination we estimated corresponding cut-off frequencies: $\omega_0/2\pi \simeq 330$ Hz and $\omega_0'/2\pi \simeq 2.6$ kHz [Fig. 2.2 (circles)]. The present and earlier experiments [59] carried out in sillenite crystals (illumination wavelength $\lambda = 442$ nm) have revealed that the values of cut-off frequencies ω_0, ω_0' are close to each other. This means that the lifetime τ and Maxwell relaxation time τ_M are comparable: $\tau \sim \tau_M \sim 100$ μs [42, 59]. In this case the cut-off frequencies are defined by the more complicated expressions than Eqs. (2.20), (2.21) [42]:

$$\omega_0 = \frac{1+\tau/\tau_I + K^2l_D^2}{\tau + \tau_M(1+\tau/\tau_I+K^2L_D^2)}, \quad (2.22)$$

$$\omega_0' = \frac{1}{\tau_M} + \frac{1+\tau/\tau_I+K^2L_D^2}{\tau} \quad (2.23)$$

$$+ \frac{1 + \tau/\tau_I + K^2 l_D^2}{\tau + \tau_M(1 + \tau/\tau_I + K^2 L_D^2)}, \tag{2.24}$$

where τ_I is the lifetime of the ionized donor state, l_D is the Debye screening length.

The condition $\tau \ll \tau_M$ is typical for the most of widegap semiconductor materials. To fulfill this condition and to conduct photocurrent measurements in the relaxation time regime additional infrared illumination of the crystal ($\lambda_{IR} = 650 - 2700$ nm, $I_{IR} = 13$ mW/mm^2) was used. For the first and second cut-off frequencies the following values were obtained: $\omega_0/2\pi \simeq 800$ Hz ($\tau_M \simeq 0.25$ ms) and $\omega_0'/2\pi \simeq 13$ kHz ($\tau \simeq 12$ μs) [Fig. 2.2 (squares)]. As seen from Fig. 2.2 the second cut-off frequency ω_0' is remarkably shifted to higher excitation frequencies what can be associated with the decrease of the electron lifetime. The growth of the first cut-off frequency ω_0 was clearly observed. This fact was unexpected since infrared illumination should not influence *stationary* photoconductivity of the crystal (and correspondingly the value of the Maxwell relaxation time). Note, that even $20-25\%$ *decay* of the sample's conductance in the presence of the infrared light was observed. For this reason the behavior of ω_0 cannot be attributed to the growth of the conductivity. The observed behavior can be explained by the fact that $\tau \sim \tau_M$. In this case the cut-off frequency depends both on the electron lifetime and the Maxwell relaxation time. The former vanishes with the additional illumination resulting in the slight change of the first cut-off frequency.

The dependencies of the corresponding cut-off frequencies versus spatial frequency of the interference pattern K are presented in Fig. 2.3. The experimental dependencies were fitted using Eqs. (1.37), (1.38) providing the following estimations for the diffusion length of photocarriers: $L_D \simeq 1.2$ μm ($\mu\tau \simeq 0.6 \times 10^{-10}$ m^2/V) [from $\omega_1(K)$], $L_D \simeq 1.6$ μm ($\mu\tau \simeq 1.0 \times 10^{-10}$ m^2/V) [from $\omega_2(K)$].

Investigation of the spatial frequency dependence of the holographic photocurrent gives another possibility for the diffusion length determination. For the diffusion mechanism of recording ($E_0 = 0$) and for modulation frequencies higher than the first characteristic cut-off frequency ($\omega > \omega_0$) the simplest theory of the effect predicts the following dependence of the photocurrent am-

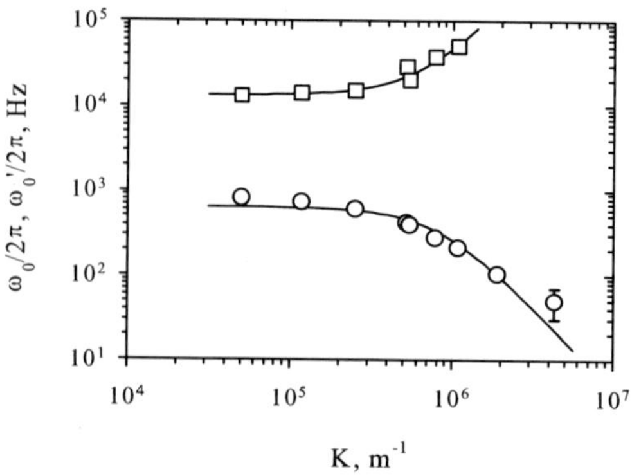

Figure 2.3. Dependence of the cut-off frequencies ω_0 (circles), ω_0' (squares) on the spatial frequency K of the interference pattern ($Bi_{12}SiO_{20}$, $\lambda = 442$ nm, $E_0 = 0$, $P_0 = 0.32$ mW, $m = 0.65$, infrared illumination is switched on). The theoretical curves are calculated using Eqs. (2.20), (2.21).

plitude versus spatial frequency [20]:

$$j^{\omega}(K) \propto \frac{K}{1 + K^2 L_D^2}. \tag{2.25}$$

We performed these measurements (Fig. 2.4) and obtained the following estimation for the diffusion length of electrons: $L_D = 1.4\ \mu$m ($\mu\tau \simeq 0.7 \times 10^{-10}$ m^2/V).

Let us proceed to the case of non-zero alternating electric field applied to the crystal. The frequency dependence of the photocurrent amplitude measured for different values of electric field applied to the crystal is presented in Fig. 2.5. As seen from this Figure all dependencies keep the form typical for the diffusion mechanism of the photocurrent excitation. No resonant peaks on the frequency response were observed even for the highest voltage applied. The giant enhancement of the signal amplitude is another feature of the nonstationary photocurrent excitation in an external AC electric field (Figs. 2.5, 2.6). The characteristic growth of the signal amplitude as $\propto E_0^2$ was observed in the region $E_0 = (0.2 - 2) \times 10^5$ V/m. The photocurrent amplification by a factor of $G_m \simeq 350$ (~ 50 dB) has been achieved for low spatial frequencies of the

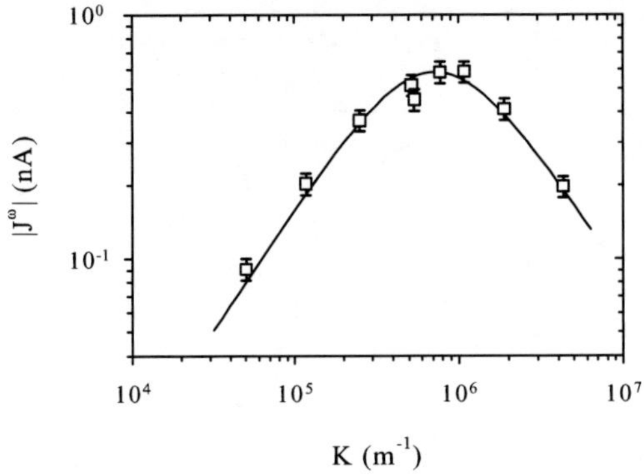

Figure 2.4. Dependence of the photocurrent amplitude $|J^{\omega}|$ on the spatial frequency K ($Bi_{12}SiO_{20}$, $\lambda = 442$ nm, $E_0 = 0$, $P_0 = 0.32$ mW, $m = 0.65$, infrared illumination is switched on). The solid line shows the theoretical curve calculated using Eq. (2.25).

interference pattern ($K^2L_D^2 \ll 1$). This value is even higher than the theoretical estimation of the maximum gain factor by Eq. (2.19): $G_m = 140 - 280$ ($K = 5 \times 10^4$ m^{-1}, $L_D = 1.2 - 1.7$ μm).

The $\mu\tau$-product or diffusion length of photocarriers can be estimated from the both characteristic breakpoints on the dependence of the photocurrent amplitude versus AC field amplitude (Fig. 2.6). We have approximated this dependence by Eq. (2.17) and obtained the following values: $E_L \simeq 1.5 \times 10^4$ V/m ($L_D \simeq 1.7$ μm, $\mu\tau \simeq 1.2 \times 10^{-10}$ m^2/V) for the first breakpoint and $E_M \simeq 2.6 \times 10^5$ V/m ($L_D \simeq 1.4$ μm, $\mu\tau \simeq 0.8 \times 10^{-10}$ m^2/V) for the second breakpoint.

The behavior of the the first cut-off frequency ω_0 versus applied electric field is similar to the drift mechanism of recording – it shifts towards low excitation frequencies (Fig. 2.7). Approximation of this dependence by Eq. (2.18) provides the value of $\mu\tau$-product: $\mu\tau \simeq 1.2 \times 10^{-10}$ m^2/V ($E_M \simeq 1.6 \times 10^5$ V/m).

The utilization of AC field for the photocurrent signal enhancement provides several important advantages over the application of DC field. First, the flat frequency response of the effect seems to be more convenient for measurements

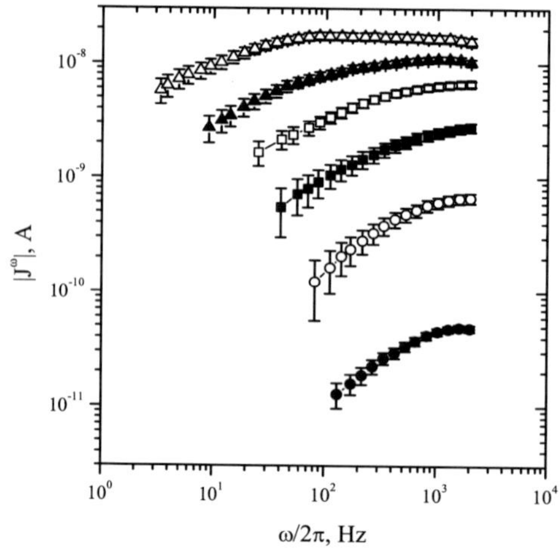

Figure 2.5. Frequency transfer functions of the photocurrent measured for $E_0 = 0$ (filled circles), 53 V/mm (open circles), 130 V/mm (filled squares), 210 V/mm (open squares), 410 V/mm (filled triangles), and 1000 V/mm (open triangles). $Bi_{12}SiO_{20}$, $\lambda = 442$ nm, $P_0 = 0.19$ mW, $m = 0.65$, $K = 5 \times 10^4$ m^{-1}, infrared illumination is switched on.

of phase modulated optical signals of small amplitude. The high frequency alternating electric field ($\Omega \gg \tau_M^{-1}$) applied to the crystal is more homogeneous than the DC field (no screening effects are presented) [44]. For this reason the measurements performed in fast AC fields are expected to be more plausible.

Determination of $\mu\tau$-product from the dependence of photocurrent amplitude versus applied AC voltage is much easier than the standard technique based on the measurements of the photocurrent amplitude on the spatial frequency of the interference pattern which is time consuming operation and includes numerous mechanical movements and readjustments of optical components (see, for example, Ref. [20]). The position of the first breakpoint on the dependence $J^{\omega}(E_0)$, i.e., E_L does not depend both on spatial frequency of the interference pattern K (for $K^2L_D^2 \ll 1$) and average light intensity I_0, what minimizes corresponding experimental errors.

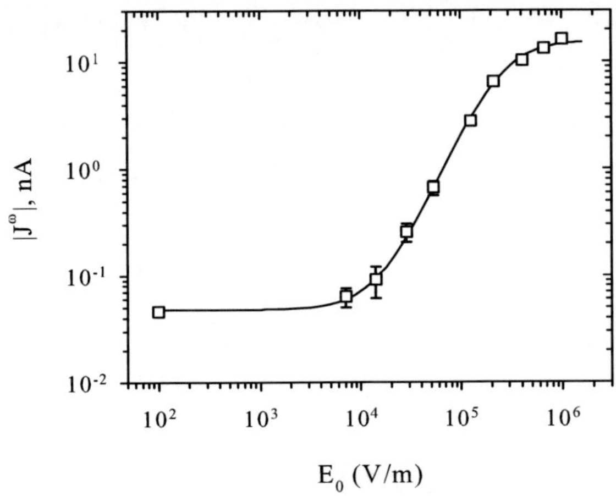

Figure 2.6. Photocurrent amplitude $|J^{\omega}|$ $(\omega > \omega_0)$ versus effective value of the external AC field E_0 ($Bi_{12}SiO_{20}$, $\lambda = 442$ nm, $P_0 = 0.19$ mW, $m = 0.65$, $K = 5 \times 10^4$ m^{-1}, $\omega/2\pi = 2.0$ kHz, infrared illumination is switched on). The measurement performed at zero field is shown at $E_0 = 10^2$ V/m. The solid line shows the theoretical curve calculated using Eq. (2.17).

2. Resonant Case

Let us consider the resonant high-frequency excitation of the non-steady-state photocurrents in photoconductive crystals placed in a sinusoidal electric field ($\omega > \Omega$, $\tau_M^{-1} < \Omega < \tau$). Such combination of frequencies can give an unique opportunity for the signal magnification. Indeed the frequency of external field is higher than the inverse formation time of the space charge grating with period $\Lambda = 2\pi/K$: $\Omega > |\tau_M(1 + K^2L_D^2 + iKL_0)|^{-1}$, so the space charge amplitude will be increased as it was predicted in Ref. [55]. At the same time the frequency of the external field is lower than the inverse relaxation time of conductivity grating: $\Omega < |(1 + K^2L_D^2 + iKL_0)\tau^{-1}|$. For this reason the amplitude of photoconductivity grating will be increased as well via resonant mechanism analogous to discussed in Ref. [38]. As a result the resonant signal enhancement in an ac electric field is expected to be more pronounced than for the case of the dc field.

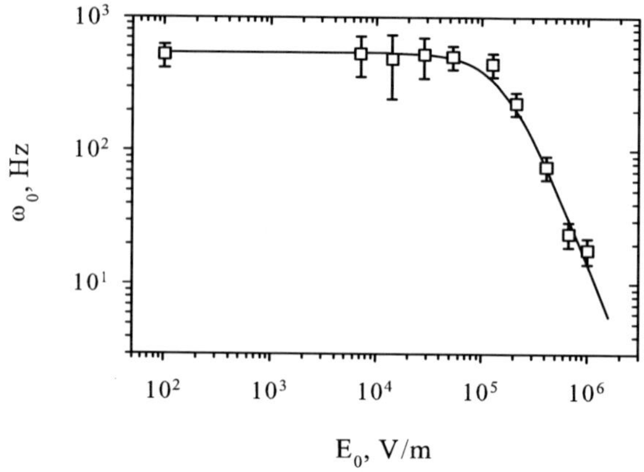

Figure 2.7. Dependence of the cut-off frequency ω_0 on the effective value of the external AC field E_0 ($Bi_{12}SiO_{20}$, $\lambda = 442$ nm, $P_0 = 0.19$ mW, $m = 0.65$, $K = 5 \times 10^4$ m^{-1}, infrared illumination is switched on). The measurement performed at zero field is shown at $E_0 = 10^2$ V/m. The approximation by Eq. (2.18) is shown by solid line.

The crystal is illuminated with the oscillating interference pattern formed by two plane light waves one of which is phase modulated with frequency ω and amplitude Δ. The sinusoidal electric field with amplitude E_{ext} and frequency Ω is applied to the crystal: $E_{ext}(t) = E_{ext} \cos \Omega t$. For the external field we use the effective value $E_0 = E_{ext}/\sqrt{2}$ rather than amplitude E_{ext}. We consider the case of "relaxation type" semiconductor , i.e. $\tau \ll \tau_M$. The opposite situation (i.e. $\tau \geq \tau_M$) is slightly more complicated for the analysis, moreover the application of strong electric fields decreases the time constant of the conductivity relaxation and increase the buildup time for the space charge, so we obtain conditions analogous to the "relaxation type" regime.

We suppose also that the frequency of phase modulation ω is much higher than the frequency of the external electric field Ω; the period of the ac field is larger than the carrier lifetime τ and shorter than grating relaxation time τ_M:

$$\omega \gg \Omega, \tag{2.26}$$

$$\tau_M^{-1} \ll \Omega \ll \tau^{-1}. \tag{2.27}$$

The calculation procedure includes the solution of the balance, continuity, and Poisson equations. Since the main stages of this procedure are similar to the ones described above we omit them and write the final expression for the complex amplitude of the photocurrent with frequency ω:

$$\begin{aligned} j^{\omega} &= -\frac{m^2\Delta}{2}\sigma_0\left(E_D + E_0\frac{KL_0}{1+K^2L_D^2}\right) \\ &\times\frac{1+K^2L_D^2+i\omega\tau}{(1+K^2L_D^2+i\omega\tau)^2+K^2L_0^2}. \end{aligned} \tag{2.28}$$

Here $E_D = (k_BT/e)K$ is the diffusion field and $L_0 = \mu\tau E_0$ is the drift length of photocarriers, k_B is the Boltzmann constant, T is the temperature. The resonant maximum appears in the frequency transfer function of the photocurrent when the drift item in this expression dominates over the diffusion one: $E_0 \gg E_D$, $KL_0 \gg 1 + K^2L_D^2$. Then the resonant frequency and corresponding signal amplitude are equal to

$$\omega_r = \frac{KL_0}{\tau} = K\mu E_0, \tag{2.29}$$

$$j^{\omega}(\omega_r) = -\frac{m^2\Delta\sigma_0E_0KL_0}{4(1+K^2L_D^2)^2}. \tag{2.30}$$

Let us compare the photocurrent excitation in the dc and ac fields for the considered frequency region: $\omega \gg \tau_M^{-1}$. The photocurrent amplitude in dc electric field was found to be [38]:

$$j_{dc}^{\omega} = \frac{m^2\Delta}{2}\sigma_0\frac{E_0KL_0 - E_D(1+K^2L_D^2+i\omega\tau)}{(1+K^2L_D^2+i\omega\tau)^2+K^2L_0^2}. \tag{2.31}$$

As seen the frequency transfer functions of the photocurrent for ac and dc fields are similar: the resonant peaks appear at the same frequency [Eq. (2.29)]. However there are distinctive features which differ these approaches. The resonant amplitude of the photocurrent excited in dc field equals

$$j_{dc}^{\omega}(\omega_r) = -i\frac{m^2\Delta\sigma_0E_0}{4(1+K^2L_D^2)}. \tag{2.32}$$

It is $KL_0/(1+K^2L_D^2)$ times lower than those for ac field [Eq. (2.30)]. Moreover this difference grows when the amplitude of applied field increases. Such advantage of the ac voltage over dc one is associated with the fact that ac field enhances the stationary space charge grating proportionally to $\propto E_0^2$ (while dc fiels does $\propto E_0$) [55]. The phases of the photocurrent amplitude defined by Eqs. (2.30) and (2.32) differ by $\pi/2$ what is due to the fact that the spatial shift of the stationary electric field grating is equal to $-\pi$ for the dc field and $-\pi/2$ for the ac field applied. [2] Besides one can note the different shapes of the resonant peak for the cases of ac and dc external field [compare Eqs. (2.28) and (2.31)].

The measurements of the non-steady-state photocurrents under applied alternating electric field were carried out using the conventional experimental setup discussed above. The second harmonic of a Nd:YAG laser ($\lambda = 532$ nm) with an average power of $P_{out} \approx 50$ mW was used as the basic source of coherent radiation. A conventional Twyman-Green interferometer forms the interference pattern with a specified fringe spacing. The contrast of the interference pattern was settled to be small enough ($m = 0.12$) to avoid the excitation of higher spatial harmonics of the electric field grating. An electrooptic phase modulator ML-102A produced phase modulation of the laser beam with frequency ω and amplitude $\Delta = 0.81$. The additional low frequency phase modulation with frequency 100 Hz and amplitude 2.4 rad was utilized for separation of the volume and contact signals. This technique allows to transfer the contact component of the signal to the side frequencies which are rejected by the selective receiver. As a result the signal detected at carrier frequency contains the volume component only. The photocurrent signal arising in the crystal's volume was amplified, filtered and then measured by spectrum analyzer SK4-58 ($f = 0.4 - 600$ kHz, $\Delta f = 100$ Hz, with 10 Hz videofilter). In our experiments we used n-type $Bi_{12}SiO_{20}$ crystal with characteristic dimensions $1 \times 4 \times 10$ mm^3, the front and back surfaces of the crystal (1×10 mm^2) were polished, the silver paste electrodes (4×3 mm^2) were painted on the lateral surfaces. The interelectrode spacing was $L = 1$ mm. The crystal was placed in the teflon holder with styrofoam linings which damp the mechanical oscillations due to the piezoelectric effect. The sinusoidal voltage from generator ($\Omega/2\pi = 6$ kHz) was amplified by conventional transformer and then applied to the sample. In some experiments the crystal was simultaneously illuminated by the light from the microscope lamp focused by lens and passed through the different filters: IKS-5 ($\lambda_{IR} = 980 - 3000$ nm), KS-15 ($\lambda_R = 650 - 2700$ nm), SZS-22 ($\lambda_B = 380 - 530$ nm). The crystal's temperature was measured by the

PtRh-Pt thermocouple.

Figure 2.8 presents the frequency transfer functions of the nonstationary holographic photocurrent measured for different amplitudes of external electric field $E_0 = U_0/L$ (U_0 is the effective amplitude of applied voltage). The resonant maximum appears in the investigated frequency region ($\omega/2\pi = 60 - 600$ kHz) when the effective value of the sinusoidal field reaches $\sim$ 5 kV/cm. The further increase of the applied voltage leads to the nonlinear growth of the resonant frequency and current amplitude (Figs. 2.8, 2.9).

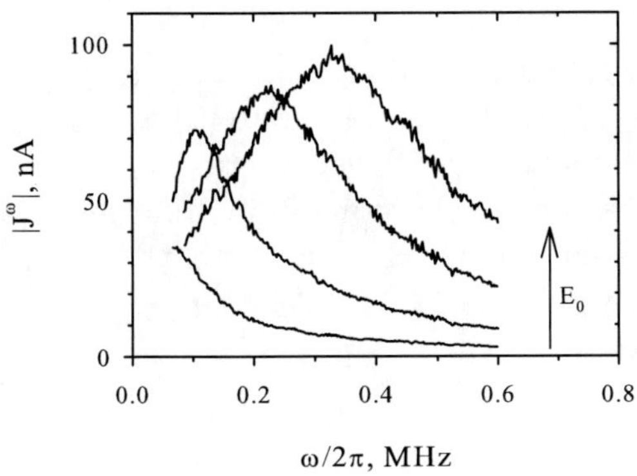

Figure 2.8. Frequency transfer functions of the holographic current measured in photorefractive $Bi_{12}SiO_{20}$ for different effective values of the sinusoidal electric field: $E_0 = 4.0, 5.8, 8.0, 9.5$ kV/cm ($\lambda = 532$ nm, $I_0 = 2.2$ W/cm^2, $K = 500$ cm^{-1}, $\Omega/2\pi = 6$ kHz).

The photocurrent amplitude grows slower than E_0^2 what is probably due to nonlinear effects. The stationary electric field grating has amplitude $\sim mE_0KL_0$ [60]. When this value approaches the amplitude of external field $E_{ext} = \sqrt{2}E_0$ we can expect the saturation of the photocurrent. For the chosen m, K and for $\mu\tau \sim 10^{-6}$ cm^2/V [60] the discussed effect should appear at $E_0 \sim 2$ kV/cm.

Nonlinear dependence $\omega_r(E_0)$ points out to the fact that strong electric fields influence the electron mobility varying in the range $\mu = 0.19 - 0.43$ cm^2/V s (Fig. 2.9). Dependence $\mu(E_0)$ was approximated with the power function: $\mu \propto E_0^{1.3}$. Such behavior is most probably explained by heating of the sample

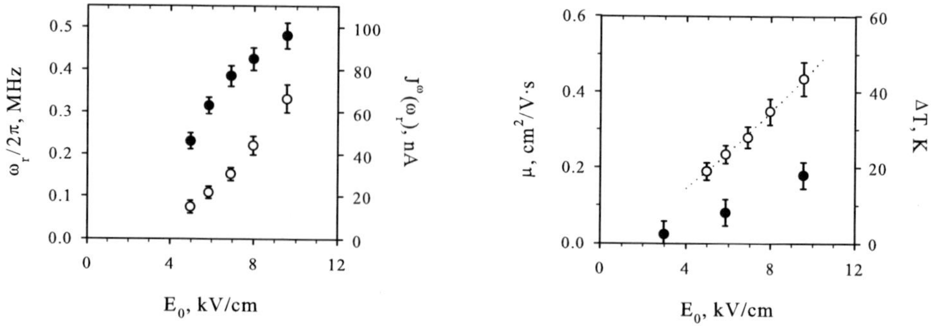

Figure 2.9. Resonant frequency ω_r (○) and resonant photocurrent amplitude $J^\omega(\omega_r)$ (●) versus effective value of the external field E_0 (left figure) and dependence of the drift mobility μ (○) versus effective value of the external field E_0 (right figure). The dotted line shows fitting function $\mu \propto E_0^{1.3}$. The increment of the crystal's temperature $\Delta T = T - 293$ K versus the external field amplitude E_0 is shown as well (●). $\lambda = 532$ nm, $I_0 = 2.2$ W/cm^2, $K = 500$ cm^{-1}, $\Omega/2\pi = 6$ kHz.

in the electric field. Indeed, the crystal's temperature raises by ~ 20 K in the external field of 9.5 kV/cm (Fig. 2.9).

The forbidden gap of sillenite type crystals is characterized by a rather complicated structure of local levels [2]. In particular, the shallow traps affect the charge transport significantly magnifying the conductivity relaxation time and reducing the effective drift mobility of electrons [44]. When the temperature grows electrons spend less time being captured on shallow traps so the effective mobility increases.

The dependencies of the resonant frequency and photocurrent amplitude versus spatial frequency of the interference pattern are presented in Fig. 2.10. Dependence $\omega_r(K)$ is well fitted with the linear function as predicted by Eq. (2.29). This approximation allows us to estimate electron mobility: $\mu = 0.42$ cm^2/V s ($T \simeq 313$ K). This value is more than one order of magnitude higher than the ones measured earlier by the non-steady-state photocurrent technique in an external dc electric field [38, 61]. In contrast to the electric field the spatial frequency of the interference pattern does not influence the photoelectric parameters (lifetime, mobility, etc.) so the behavior of the resonant frequency

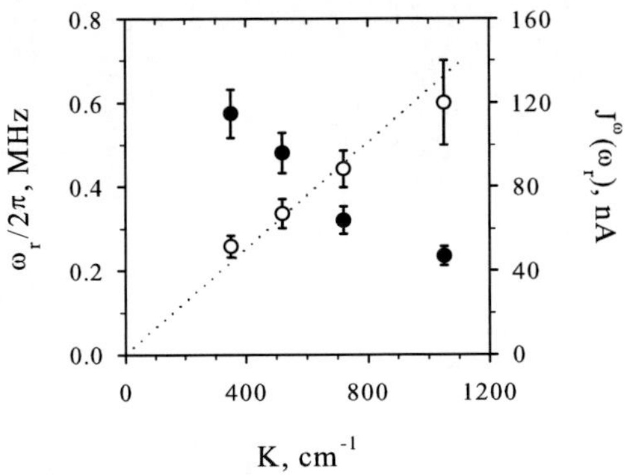

Figure 2.10. Dependencies of the resonant frequency ω_r (○) and resonant photocurrent amplitude $J^{\omega}(\omega_r)$ (●) on the spatial frequency of the interference pattern K ($\lambda = 532$ nm, $I_0 = 2.2$ W/cm^2, $E_0 = 9.5$ kV/cm, $\Omega/2\pi = 6$ kHz, $T \simeq 313$ K). The dotted line shows the approximation by Eq. (2.29) for $\mu = 0.42$ cm^2/V s.

are in good agreement with the simplest theory described above and we can state that the maximum on the frequency transfer functions of the photocurrent is associated with the resonant excitation of the photoconductivity gratings.

As seen from Fig. 2.10 the photocurrent amplitude is maximal for the lowest settled spatial frequency $K = 3 \times 10^2$ cm^{-1} though the theoretical estimation by Eq. (2.30) predicts maximum at $K = (\sqrt{3}L_D)^{-1} \sim 6 \times 10^3$ cm^{-1} ($L_D \sim 1$ μm). Perhaps this discrepancy is typical for sillenites and associated with the grating saturation effect discussed above.

Dependencies of the resonant frequency and current amplitude versus light intensity are shown in Fig. 2.11. According to the basic theory presented above no dependence of the resonant frequency ω_r should be observed [Eq. 2.29]. In practice, however, this statement is not valid since light intensity can influence the effective mobility via at least two mechanisms. First, the mobility increases due to Joule heating of the illuminated crystal placed in the electric field. Second, the increase of the light intensity changes the population of the shallow traps, the capture rate of electrons to these traps becomes weaker what consequently increases the effective mobility. As a result we obtained the wide range

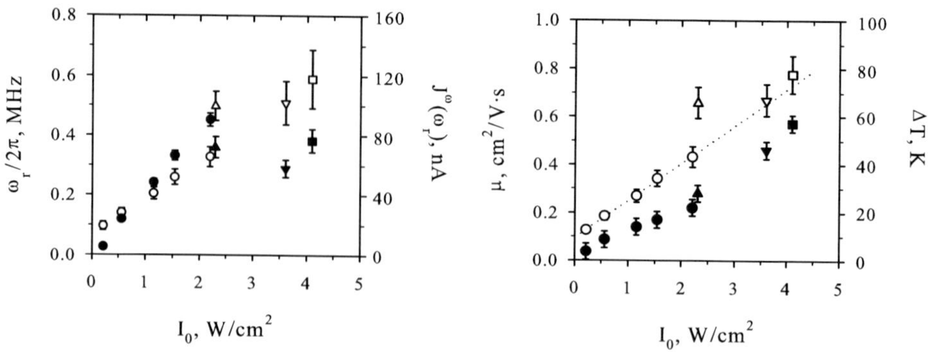

Figure 2.11. Resonant frequency ω_r (○, △, ▽, □) and resonant photocurrent amplitude $J^\omega(\omega_r)$ (●, ▲, ▼, ■) versus total light intensity I_0 (left figure) and dependence of the drift mobility μ (○, △, ▽, □) on the total light intensity I_0 (right figure). Dependence $\mu(I_0)$ measured without additional illumination (○) is fitted by the linear function $\mu \propto I_0$ with $\mu(0) = 0.95$ cm^2/V s. Dependence of the temperature increment $\Delta T = T - 293$ K on the light intensity I_0 is shown by the filled symbols (●, ▲, ▼, ■) . The circles (●, ○) demonstrate the measurements with the only laser illumination ($\lambda = 532$ nm), the other symbols correspond to the measurements with the additional illumination: ▲, △ – $\lambda_B = 380 - 530$ nm, $I_B = 0.10$ W/cm^2; ▼, ▽ – $\lambda_{IR} = 980 - 3000$ nm, $I_{IR} = 1.4$ W/cm^2; ■, □ – $\lambda_R = 650 - 2700$ nm, $I_R = 1.9$ W/cm^2. $K = 500$ cm^{-1}, $E_0 = 9.5$ kV/cm, $\Omega/2\pi = 6$ kHz.

for electron mobility: $\mu = 0.13 - 0.43$ cm^2/V s. The similar behavior of the photocurrent was observed earlier for the case of dc external field [61].

In order to vary the total light intensity and spectrum the crystal was illuminated by the additional infrared, red, blue and "white" light from the microscope lamp (Fig. 2.11). The most powerful "white" light with the intensity $I_W = 2.9$ W/cm^2 shifts the resonant maximum beyond the studied frequency range ($\omega_r/2\pi > 600$ kHz) what correspond to the value of drift mobility $\mu > 0.8$ cm^2/V s ($T \simeq 370$ K).

The dependence of the resonant peak amplitude on the light intensity ($\lambda = 532$ nm) was found to be linear (Fig. 2.11) what is associated with the linear character of the photoconductivity. As seen from the same figure the additional illumination suppress the signal amplitude. Two reasons of this phenomena

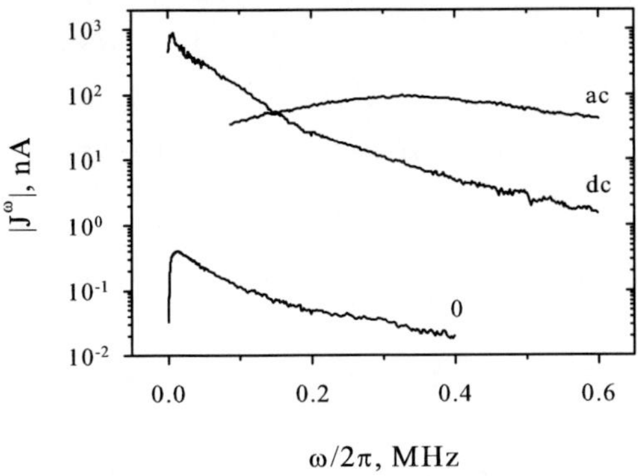

Figure 2.12. Frequency transfer functions of the holographic photocurrent measured in ac ($E_0 = 9.5$ kV/cm, $\Omega/2\pi = 6$ kHz, $T \simeq 313$ K), dc ($E_0 = 10$ kV/cm, $T \simeq 313$ K), and zero ($E_0 = 0$, $T \simeq 293$ K) external field. $\lambda = 532$ nm, $I_0 = 2.2$ W/cm^2, $K = 500$ cm^{-1}.

can be adduced. First, the incoherent backlight decreases the contrast of the interference pattern and photocurrent amplitude as a consequence. Second, the simultaneous infrared illumination redistributes the electrons between different types of local levels reducing the effective lifetime, nonstationary response of the photoconductivity, and resulting photocurrent [44,60].

We have compared two techniques for the drift mobility determination and carried out measurements of the frequency transfer functions of the nonstationary photocurrent excited in ac and dc fields with approximately equal effective values $E_0 \simeq 10$ kV/cm (Fig. 2.12). As follows from this figure the photocurrent generation in ac field is more efficient in the high frequency region $\omega > 300$ kHz: the signal enhancement with respect to diffusion regime equals ~ 70 dB for the case of ac field and ~ 50 dB for dc one. This fact proves the theoretical estimations [Eqs. (2.30), (2.32)] that predicts the prevalence of ac field excitation over dc one by a factor $KL_0 \sim 10$.

One can notice the most interesting difference between these two approaches: the frequency transfer function of the nonstationary photocurrent excited in dc field contains no resonant maximum at high modulation frequencies $\omega > \omega_0'$ [where $\omega_0'/2\pi \simeq 20$ kHz is the second cut-off frequency of the nonstationary photocurrent excited without an external field (Fig. 2.12)].

It is quite difficult to point out the origin of this feature. (i) First the silver paste contacts to $Bi_{12}SiO_{20}$ crystals are usually rectifying so the 10-20% dc voltage drop in the contact area is expected. This decreases KL_0-product partially eliminating the resonant conditions for the dc field. The blocking contact gives rise to the inhomogeneous distribution of dc field over the interelectrode spacing as well. It makes the resonant frequency to be dependent on the local value of the electric field what can results in blurring of the resonant peak. In the case of high frequency ac field ($\Omega > \tau_M^{-1}$) such effects are negligible as the displacement current through the contacts decreases corresponding voltage drops and makes the electric field more homogeneous.

(ii) Another and perhaps the most essential reason for the discussed difference between photocurrent generation in ac and dc fields can be associated with the complicated processes of conductivity relaxation in $Bi_{12}SiO_{20}$. The frequency dependence of the photoconductivity response to the amplitude modulated light [44] does not adequately described by the simplest band transport model with one type of partially compensated donor level. Namely there are at least two regions where photoconductivity relaxation is characterized by different time constants (Fig. 2.13). In our experiments the frequency of the ac field ($\Omega/2\pi = 6$ kHz) was higher than the characteristic frequency separating these regions ($\omega_s/2\pi \sim 100$ Hz). We suppose that frequency transfer functions measured in ac and dc fields differ because the effective photoelectric parameters governing the charge transport does not coincide for these cases: $\tau_{ac} < \tau_{dc}$ and $\mu_{ac} > \mu_{dc}$. This assumption is proved by the values of effective mobility measured by the discussed technique in ac field $\mu_{ac} = 0.13 - 0.8$ cm^2/V s and dc one $\mu_{dc} = 0.016 - 0.043$ cm^2/V s [38, 61]. So the larger mobility in ac field helps to fulfill resonant condition $\omega_r \gg \omega_0'$ and obtain the resonant maximum on the frequency dependence of the photocurrent.

The resonant excitation of the non-steady-state photocurrents in an ac field has demonstrated several distinctive features that can improve the techniques and devices based on the discussed effect. In particular, the proposed above resonant technique allows the direct measurements of the photocarrier's drift mobility. The utilization of an ac external voltage instead of the dc one minimizes the experimental errors of the mobility measurements since the voltage drop in the contact area is smaller and the electric field distribution in the sample

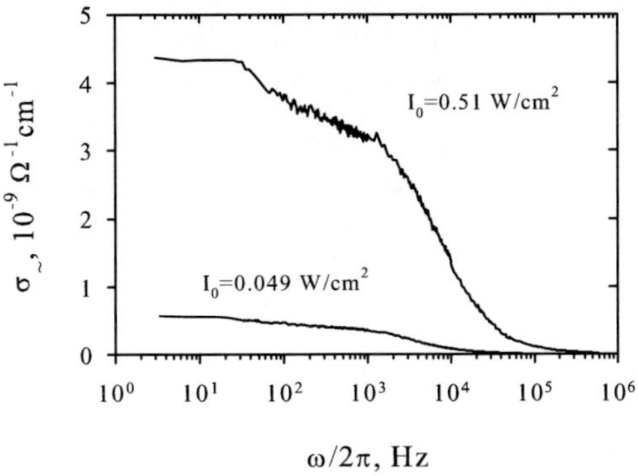

Figure 2.13. Frequency dependence of the nonstationary photoconductivity response to the amplitude modulated light with different average intensities $I_0 = 0.049, 0.51$ W/cm^2. $\lambda = 532$ nm, $m = 0.018$.

volume is more homogeneous. Besides, variations of the voltage frequency may provide additional possibilities for the characterization of crystals with complicated conductivity relaxation processes.

The drift mobility of electrons in $Bi_{12}SiO_{20}$ was measured using this method. It lies in the range $\mu = 0.13 - 0.8$ cm^2/V s which is approximately an order lower than the range of actual mobility $\mu = 3.4 - 5.5$ cm^2/V s estimated by other techniques. On the other hand the measured values are $3 - 10$ times higher than the ones obtained by the non-steady-state photo-EMF method realized in dc fields [38, 61]. So the application of an ac electric field allowed us to partially overcome difficulties of the mobility measurements in widegap semiconductors with shallow energy levels.

The giant signal enhancement in the high frequency region is another attractive feature of the photocurrent generation in an alternating field. The enhancement factor reaches $\sim$ 70 dB for low spatial frequencies of the interference pattern against the factor of $\sim$ 50 dB typical for the case of dc field.

For practical applications it is more useful to calculate and compare the signal-to-noise ratio for zero and ac fields. Let us estimate it for the resonant frequency ($\omega = \omega_r \gg \omega_0'$), low spatial frequency ($K^2 L_D^2 \ll 1$), and low load resistance ($R_L \ll Z_{cr}$, Z_{cr} is the crystal's impedance). For the mentioned frequency region the signal amplitude can be calculated using Eq. (2.28). The thermal noise of the load resistor is dominant for the diffusion regime of the photocurrent generation: $J_{th} = (4k_B T R_L^{-1} \Delta f)^{1/2}$ (Δf is the detection bandwidth). If the external voltage is applied then generation-recombination noise can be prevalent: $J_{g/r} = 2e\mu\tau E_0 [S g_0 \Delta f / L(1+\omega^2\tau^2)]^{1/2}$ (g_0 is the average generation rate of photoelectrons, S is the cross section of the illuminated volume) [62]. Finally the magnification of the signal-to-noise ratio equals

$$\frac{(S/N)_{ac}}{(S/N)_0} = \frac{E_0 K^2 L_0^2}{2} \sqrt{\frac{L}{S g_0 k_B T R_L}} \quad \text{for } J_{g/r} \gg J_{th} \tag{2.33}$$

or

$$\frac{(S/N)_{ac}}{(S/N)_0} = \frac{e E_0 K L_0^2}{2 k_B T} \quad \text{for } J_{g/r} \ll J_{th}. \tag{2.34}$$

This gain factor is very large: for the sensible experimental conditions and crystal parameters ($E_0 = 10$ kV/cm, $K = 500$ cm^{-1}, $L_0 = 0.02$ cm, $L = 0.1$ cm, $S = 10^{-2}$ cm^2, $g_0 = 10^{19}$ cm^{-3}s^{-1}, $R_L = 10^4$ Ω) it reaches 90 dB. In practice, however, some factors can reduce this striking value: nonlinear recording of the space charge grating, trap saturation, etc.

3. Combined Excitation of Running Space Charge and Conductivity Gratings

The evolution of photoinduced charge in semiconductor materials mainly occurs in two basic stages: formation of the photoconductivity distribution and creation of the space charge [2, 44]. For the crystal placed in an external electric field the relaxation of photoconductivity and space charge distributions has complex behavior (oscillation and running character) [5]. The existence of two eigenmodes (branches) of space charge oscillation was predicted in Ref. [48]. One of these eigenmodes describe oscillation of the space charge trapped on the local levels, while the other is associated with the oscillation of the density of electrons in the conduction band (conduction electrons). The former branch is historically called “space charge wave”. To avoid the complexity we shall use the term “running grating of the conductivity” for the latter one. The dispersion laws of these waves, i.e. dependencies of the temporal oscillation frequency on the spatial frequency K can be written as follows:

$$\omega_{sc} = (\tau_M K L_0)^{-1}, \quad (2.35)$$

$$\omega_{pc} = K\mu E_0, \quad (2.36)$$

where τ_M is the Maxwell relaxation time, μ is the mobility, and L_0 is the drift length of electrons in dc electric field E_0. A number of approaches for investigations of the running space charge and conductivity gratings has been realized by the present time [20, 32, 37, 38, 50, 63–69].

In spite of the large number of publications dedicated to the subject of space charge waves the question of simultaneous excitation and interaction of the eigenmodes mentioned above is still unsolved. Most of the proposed optical techniques use the interference pattern with the specified spatial frequency K and temporal frequency ω for the excitation of either space charge wave or running photoconductivity grating. This is due to the fact that the eigenfrequency of space charge oscillations in high resistive crystals (such as $Bi_{12}SiO_{20}$) is in the range $\omega_{sc}/2\pi = 10 - 100$ Hz while the eigenfrequency of the photoconductivity grating with the same spatial period can reach considerably higher values: $\omega_{pc}/2\pi = 0.01 - 1$ MHz. The situation is partly similar to the light scattering on the acoustic waves where interaction of particles with comparable momenta (wave vectors) and completely different energies (frequencies) takes place.

As it is known from the oscillation theory, the behavior of systems with two degrees of freedom have distinctive features with respect to the one in systems with one degree of freedom. This features result in such effects as incoherent parametric amplification, generation, wideband parametric frequency division, etc. Considering two eigenmodes of space charge oscillations in photorefractive media one can expect some analogs in holographic recording.

For the realization of the simultaneous excitation and interaction of the space charge wave (ω_{sc}, K) travelling along the dc electric field with the velocity ω_{sc}/K and electron conductivity grating (ω_{pc}, $-K$) running in the opposite direction with velocity $-\omega_{pc}/K$ the third oscillation $(\Omega, 0)$ satisfying the conditions analogous to the energy and momentum conservation laws is required: $\omega_{sc} + \omega_{pc} = \Omega$, $K - K = 0$. The electric field with frequency Ω can play the role of such oscillation and provide necessary coupling.

In this section we consider one of the possible manifestation of this phenomenon, namely the excitation of the space charge wave provided by the interaction of the running photoconductivity grating with an applied ac electric field. The effect is observed and investigated using optical and electrical methods [70–72]. The first one is based on light diffraction on photorefractive volume holograms, the second one uses the non-steady-state photoelectromotive force effect.

Let us suppose the photorefractive crystal placed in external dc and ac electric fields is illuminated by the interference pattern formed by two plain light waves one of which is phase modulated with amplitude Δ and frequency ω [Fig. 2.14, Eq. (1.6)]. The external dc and sinusoidal ac voltages applied to the crystal can be presented as follows:

$$U_{ext} = U_0 + U_A \cos \Omega t. \tag{2.37}$$

We consider the widely used model of semiconductor with monopolar photoconductivity and one type of partially compensated donor levels (Fig. 1.8) [2, 44]. The formulation of the problem and initial stages are similar to those presented in Section 1.: we have to solve Eqs. (2.6), (2.7) and find distributions of the electron density $n(x,t)$ and electric field $E(x,t)$ in the form of stationary and running sinusoidal gratings [see Eqs. (2.8), (2.9)]. The right hand of Eq. (2.9) should also contain the dc component of the electric field $E_0 = U_0/L$ in the present analysis.

Let us analyze the case of "relaxation type" semiconductor:

$$\tau \ll \tau_M. \tag{2.38}$$

Note, that the opposite case (i.e. $\tau \geq \tau_M$) is slightly more complicated for the analysis. However, the application of strong electric fields decreases the time constant of the conductivity relaxation and increases the buildup time of the space charge, so we obtain conditions analogous to the "relaxation type" regime.

Let us consider the "long-drift length" regime of holographic recording:

$$KL_0 \gg 1 + K^2 L_D^2, \tag{2.39}$$

where $L_0 = \mu\tau E_0$ and $L_D = (k_B T \mu\tau/e)^{1/2}$ are the drift and diffusion lengths of photocarriers. This assumption is rather typical for theoretical considerations

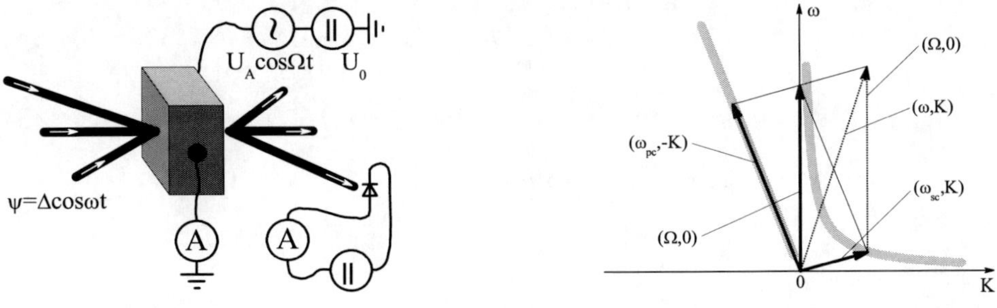

Figure 2.14. Scheme for observation of the combined excitation of running space charge and conductivity gratings (left figure). Diagram illustrating the interaction of running space charge and conductivity gratings with applied ac electric field. The arrows show the space charge grating (ω_{sc}, K) travelling with the velocity ω_{sc}/K along the dc field, the photoconductivity grating $(\omega_{pc}, -K)$ running with the velocity $-\omega_{pc}/K$ in the opposite direction, the photoconductivity grating (ω, K) driven with the velocity ω/K along the dc field, and the external ac field $(\Omega, 0)$ with frequency Ω. The half-tone lines show the dispersion laws for the space charge wave ω_{sc}(K) and running photoconductivity grating $\omega_{pc}(-K)$ (right figure).

of various phenomena in photorefractive media: it ensures the oscillatory relaxation of the space charge and photoconductivity for large applied dc fields (in contrast to the exponential one in the diffusion regime) [2].

In this section we shall develop two approaches. In the former one the frequency of an ac electric field is equal to the eigenfrequency of photoconductivity oscillation, and the difference frequency lies in the range of the space charge wave excitation:

$$\Omega = \omega_{pc}, \tag{2.40}$$

$$|\omega - \Omega| \sim \omega_{sc}. \tag{2.41}$$

The complex amplitudes of values $n(x,t)$ and $E(x,t)$ necessary for further analysis have been calculated. Below the most interesting terms presenting gratings with spatial frequency K and oscillation frequency $\pm(\omega - \Omega)$ are given:

$$n^{++-} = \frac{-im\Delta}{8D_1} n_0(\omega - \Omega)\tau_M K L_A, \tag{2.42}$$

$$E^{++-} = \frac{-im\Delta}{8D_1} E_A, \tag{2.43}$$

$$n^{+-+} = \frac{i3m\Delta}{8D_2} n_0(\omega - \Omega)\tau_M K L_A, \tag{2.44}$$

$$E^{+-+} = \frac{-i3m\Delta}{8D_2} E_A, \tag{2.45}$$

where

$$\begin{aligned} D_1 &= (1 + K^2 L_D^2)[1 + (\omega - \Omega)\tau_M K L_0] \\ &\quad - i[(1 + K^2 L_D^2)^2 + K^2 L_A^2/4]/K L_0, \end{aligned} \tag{2.46}$$

$$\begin{aligned} D_2 &= (1 + K^2 L_D^2)[5 - 11(\omega - \Omega)\tau_M K L_0] \\ &\quad - i6KL_0[1 - (\omega - \Omega)\tau_M K L_0], \end{aligned} \tag{2.47}$$

and $L_A = \mu\tau E_A$.

The latter approach implies that frequencies of phase modulation and external field vary in the region of the photoconductivity grating oscillation maintaining the difference frequency constant and equal to the frequency of the space charge oscillation:

$$\omega, \Omega \sim \omega_{pc}, \tag{2.48}$$

$$\omega - \Omega = -\omega_{sc}. \tag{2.49}$$

The choice of negative sign in right hand of Eq. (2.49) will be clear below when we shall analyze the frequency characteristics of the non-steady-state photo-EMF and diffraction efficiency.

The corresponding complex amplitudes of electron density and electric field are the following:

$$n^{++-} = -\frac{m\Delta}{8D_3} n_0 K L_A \left[1 + K^2 L_D^2 + i(2\omega\tau - K L_0)\right], \tag{2.50}$$

$$E^{++-} = \frac{m\Delta}{8D_3} E_0 K L_A \left[1 + K^2 L_D^2 + i(2\omega\tau - K L_0)\right], \tag{2.51}$$

where

$$\begin{aligned} D_3 &= (1 + K^2 L_D^2)\left[(1 + K^2 L_D^2)^2 + K^2 L_A^2/2 - K^2 L_0^2 + 3\omega\tau K L_0 \right. \\ &\quad \left. - 2\omega^2\tau^2\right] + i\left[(1 + K^2 L_D^2)^2(3\omega\tau - 2KL_0) + K^2 L_A^2(2\omega\tau - K L_0)/4\right]. \end{aligned} \tag{2.52}$$

Let us now proceed with the calculation of the non-steady-state photo-EMF and diffraction efficiency amplitudes – values which will be measured in the experiment. As it was shown in Ref. [20] the non-steady-state photo-EMF in crystal biased by voltage source is defined by the averaged drift component of the photocurrent [Eq. (1.17)]. As seen from Eqs. (2.8), (2.9), (1.17) the signal with frequency $\omega - \Omega$ is produced by the following combination:

$$\begin{aligned} j^{\omega-\Omega}(t) &= \mathrm{Re}\{2e\mu(n^{++-}E^{-00} + n^{-00}E^{++-} + n^{-+-}E^{+00} \\ &\quad + n^{+00}E^{-+-} + n^{++0}E^{-0-} + n^{-0-}E^{++0} \\ &\quad + n^{-+0}E^{+0-} + n^{+0-}E^{-+0})\exp[i(\omega - \Omega)t]\}. \end{aligned} \quad (2.53)$$

The different nature of the items in this expression should be pointed out. The first four components are associated with the resonant excitation of the space charge waves with temporal frequency $\omega - \Omega$. The last four components in Eq. (2.53) are just mixing of the "fast" gratings with temporal frequencies ω and Ω. It is reasonable to expect that the former group produces larger contribution to the total current. This supposition is proved by the numerical calculation. Substituting the calculated above terms (2.42)-(2.45) into the Eq. (2.53) we can write down the expression for the complex amplitude of the nonstationary photocurrent as:

$$j^{\omega-\Omega} = -i\frac{m^2\Delta}{8}\sigma_0 E_A \left[\frac{1-(\omega-\Omega)\tau_M K L_0}{D_1} - 3\frac{1+(\omega-\Omega)\tau_M K L_0}{(D_2)^*}\right]. \quad (2.54)$$

The frequency transfer function of the non-steady-state photo-EMF, i.e. dependence $j^{\omega-\Omega}(\omega - \Omega)$ has two resonant maxima at $\omega - \Omega \simeq \pm\omega_{sc}$. The amplitudes of the resonant maxima at $\omega - \Omega = -\omega_{sc}$ and at $\omega - \Omega = \omega_{sc}$ are the following:

$$j_{r1}^{\omega-\Omega} = \frac{m^2\Delta}{4}\frac{\sigma_0 E_A K L_0}{(1+K^2L_D^2)^2 + K^2L_A^2/4}, \quad (2.55)$$

$$j_{r2}^{\omega-\Omega} = -i\frac{m^2\Delta}{8}\frac{\sigma_0 E_A}{1+K^2L_D^2}. \quad (2.56)$$

The resonant maximum for the negative frequency difference is approximately $2KL_0$ times larger than for positive frequency difference. This fact is associated with the principal distinction in the physical mechanism of grating's excitation in these frequency regions.

For the negative frequency difference range ($\omega - \Omega < 0$) the space charge wave is originated from the interaction of an external field with the photoconductivity grating propagating in the direction opposite to the dc field. When the oscillations of the interference pattern are synchronized with the movement of drifting electrons the resonant enhancement of the photoconductivity grating takes place. Thus the simultaneous generation of the running space charge and photoconductivity gratings, which are the eigenwaves in this case, should be observed.

For positive frequency difference ($\omega - \Omega > 0$) the space charge wave arises due to the interaction of the external field with the conductivity grating driven in the direction of the dc field. The photoconductivity grating is not an eigenmode of semiconductor in this case. It just follows the movements of the oscillating interference pattern in the direction of dc field.

The diagram illustrating the interaction of the space charge and photoconductivity gratings with the applied ac field is presented in Fig. 2.14. As seen from the figure in the former case the applied ac field couples two eigenmodes of semiconductor satisfying the equations $\omega_{sc} + \omega_{pc} = \Omega$, $K - K = 0$. In the latter case the only eigenmode, namely, space charge wave is excited. The fulfillment of the resonant conditions for both the space charge wave and photoconductivity grating ensures observation of larger signal's amplitudes.

Dependence $j^{\omega-\Omega}(\omega)$ with the fixed frequency difference $\omega-\Omega$ [Eqs. (2.48) and (2.49)] can be measured in the experiment as well. Direct substitution of Eqs. (2.50), (2.51) into Eq. (2.53) gives the following expression for this dependence:

$$j^{\omega-\Omega} = \frac{m^2\Delta}{4D_3}\sigma_0 E_A K L_0 \left[1 + K^2 L_D^2 + i(2\omega\tau - K L_0)\right]. \qquad (2.57)$$

The analysis of the denominator in this expression [see Eq. (2.52)] revealed that the resonant maximum can be observed at $\omega \simeq \omega_{pc}$ and its amplitude is expressed by Eq. (2.55). This resonant peculiarity can be explained as follows: since the excited space charge oscillations at combination frequency are due to interaction of the external ac electric field and the photoconductivity grating, the amplitude of the latter should affect the detected signal amplitude. The photoconductivity grating is resonantly excited at $\omega = \omega_{pc}$. So the resonant dependence of the low frequency signal $j^{\omega-\Omega}$ versus phase modulation frequency ω (or the frequency of an ac electric field Ω) can be expected as well.

The diffraction efficiency of thin phase hologram [2] recorded in photorefractive crystal equals

$$\eta(t) = |\pi n_p^3 r E^+(t) d/\lambda_p|^2, \tag{2.58}$$

where n_p and r are the refractive index and electrooptic coefficient of the crystal for wavelength λ_p of the probe readout light, $E^+(t)$ is the complex amplitude of $+K$ spatial component of the electric field, and d is the thickness of the crystal. The oscillations of the diffraction efficiency with frequency $\omega - \Omega$ are produced by the following components:

$$\begin{aligned} \eta^{\omega-\Omega}(t) &= \mathrm{Re}\{2q^2[(E^{+00})^* E^{++-} + E^{+00}(E^{+-+})^* \\ &+ (E^{+0+})^* E^{++0} + E^{+0-}(E^{+-0})^*] \exp[i(\omega - \Omega)t]\}, \end{aligned} \tag{2.59}$$

here $q = \pi n_p^3 r d/\lambda_p$. One can note the difference between two first and two last terms of the expression – similar to the ones mentioned above for the non-stationary photocurrent analysis. Direct substitution of Eqs. (2.43), (2.45) into Eq. (2.59) provides the expression for the complex amplitude of the diffraction efficiency:

$$\eta^{\omega-\Omega} = i\frac{m^2\Delta}{8} q^2 E_0 E_A \left[\frac{1}{D_1} - \frac{3}{(D_2)^*}\right]. \tag{2.60}$$

This frequency transfer function has two resonant maxima. The oscillation amplitudes of the diffraction efficiency at resonant frequencies $\omega - \Omega = -\omega_{sc}$ and $\omega - \Omega = \omega_{sc}$ are the following:

$$\eta_{r1}^{\omega-\Omega} = -\frac{m^2\Delta}{8} \frac{q^2 E_0 E_A K L_0}{(1 + K^2 L_D^2)^2 + K^2 L_A^2/4}, \tag{2.61}$$

$$\eta_{r2}^{\omega-\Omega} = i\frac{m^2\Delta}{16} \frac{q^2 E_0 E_A}{1 + K^2 L_D^2}. \tag{2.62}$$

Like in Eqs. (2.55), (2.56) these resonant peaks differ by factor of $\sim 2KL_0$ due to the same reason: the simultaneous excitation of two eigenmodes is more efficient.

Dependence of the diffraction efficiency amplitude $\eta^{\omega-\Omega}$ on the frequency of phase modulation ω with fixed difference frequency $\omega - \Omega = -\omega_{sc}$ can be calculated substituting Eq. (2.51) into Eq. (2.59):

$$\eta^{\omega-\Omega} = -\frac{m^2\Delta}{8D_3}q^2 E_0^2 K L_A \left[1 + K^2 L_D^2 + i(2\omega\tau - K L_0)\right]. \qquad (2.63)$$

The resonant maximum with amplitude (2.61) is observed at frequency $\omega \simeq \omega_{pc}$.

This frequency dependence along with the signal dependence of the non-steady-state photo-EMF effect [Eq. (2.57)] can be utilized for direct measurements of the photocarrier's drift mobility $\mu = \omega_{pc}/K E_0$.

We have considered here the simplest case of thin hologram recorded in slightly absorbing media. The generalization of the theory for the cases of thick holograms in absorbing and optically active materials can be done as required using the approaches known in literature [2, 5, 73].

The experimental arrangement and procedure used for the investigation of the combined excitation of space charge and conductivity gratings in photorefractive crystals are described in Refs. [70–72].

For the determination of eigenfrequencies of space charge and conductivity oscillations let us consider the results of the diffraction measurements in photorefractive $Bi_{12}SiO_{20}$ using standard technique of signal excitation.

The diffraction efficiency of the hologram was defined as $\eta = I_{p1}/I_p$, where I_{p1} is the intensity of the probe beam for the first order of diffraction. The diffraction efficiency of the stationary hologram recorded in the electric field $E_0 = U_0/L = 16$ kV/cm was equal to $\eta_0 = 2.0\%$. Figure 2.15 presents the frequency transfer function of the diffraction efficiency oscillations η^{ω} measured at frequency of phase modulation ω. The first resonant maximum is observed in the low frequency range ($\omega/2\pi = 27 - 110$ Hz). When the electric field is increased the signal amplitude increases as well and the first maximum shifts to lower excitation frequencies (Fig. 2.15, inset). The shoulder at frequencies $\omega/2\pi = 10 - 50$ kHz was observed as well. Perhaps, for larger applied electric fields this shoulder could transfer into the second maximum, but field amplitudes $E_0 > 8$ kV/cm shift this shoulder beyond the investigated frequency range. The signal amplitude at $\omega/2\pi > 50$ kHz was so small that utilization of spectrum analyzer with appropriate frequency range and larger detection bandwidth (SK4-58) was not successful.

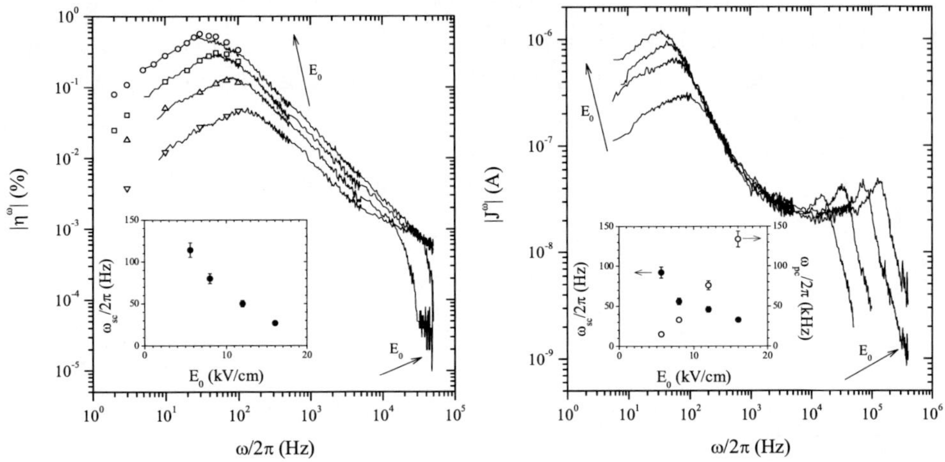

Figure 2.15. Frequency transfer functions of the diffraction efficiency oscillations (left figure) and non-steady-state photo-EMF (right figure) excited in $Bi_{12}SiO_{20}$ at the frequency of phase modulation ω (standard technique, $E_A = 0$). The dependencies are measured for four values of the applied dc electric field: $E_0 = 5.6$ kV/cm (▽), 8.0 kV/cm (△), 12 kV/cm (□), 16 kV/cm (○). The measurements performed using spectrum analyzer and lock-in voltmeter are presented by the solid lines and symbols respectively. The arrows shows the directions of curve evolution with increasing field amplitude E_0. Dependencies of the first (●) and second (○) resonant frequencies versus applied dc field are presented in the insets.

For this reason we used another approach based on the non-steady-state photo-EMF measurements. The frequency transfer function of the non-steady-state photo-EMF signal excited in $Bi_{12}SiO_{20}$ crystal is presented in Fig. 2.15. Two resonant maxima on the frequency dependencies were observed. The increase of an applied field amplitude shifts the first and second maxima to the low and high frequency bands respectively (Fig. 2.15, inset). As it was shown in Ref. [38] such signal behavior is associated with the resonant excitation of running gratings of the space charge and conductivity described by the dispersion laws (2.35) and (2.36). Note, that the first resonant maximum is wide, this can be due to nonlinearity of the space charge formation for large contrast of the

Table 2.1. Drift mobility of electrons in $Bi_{12}SiO_{20}$ crystal estimated from the frequency transfer function of the non-steady-state photo-EMF $|J^{\omega}(\omega)|$

E_0, kV/cm	T, K	μ, cm^2/V s
5.6	294	0.89×10^{-2}
8.0	296	1.4×10^{-2}
12	298	2.1×10^{-2}
16	300	2.8×10^{-2}

interference pattern [74].

The non-steady-state photo-EMF technique provides the unique opportunity for the direct determination of the photocarrier drift mobility. The value of the second resonant frequency measured for the specified K and E_0 is associated with the electron mobility value (Table 2.1). The obtained data are in good agreement with the values measured earlier [35,38,61].

The preliminary experiments allowed us to estimate corresponding resonant frequencies which can be used for further analysis of the simultaneous generation of the space charge and photoconductivity gratings.

For the realization of such regime the frequency of the external ac field was settled to be approximately equal to the frequency of the second resonant maximum. Small deviations ($0.3 - 1.0$ kHz) of frequency $\Omega/2\pi$ from the resonant frequency value $\omega_{pc}/2\pi$ are acceptable since the width of the second resonant maximum reaches $15 - 130$ kHz (see Fig. 2.15). The amplitude of an external ac electric field was chosen noticeably smaller than the value of the dc field ($E_A = U_A/L = 1.4 - 5.6$ kV/cm), so possible shifts of the resonant frequencies are expected to be negligible.

Figure 2.16 (left figure) presents the frequency transfer function of the diffraction efficiency oscillations excited at difference combination frequency $\omega - \Omega$. Two maxima were observed at this dependence: the first one – at the negative difference frequency $(\omega - \Omega)/2\pi = -(100 - 33)$ Hz ($E_0 = 5.6 - 16$ kV/cm), and the second one – at the positive frequency $(\omega - \Omega)/2\pi = 40$ Hz ($E_0 = 16$ kV/cm). The increase of the external dc field leads to the growth of the signal amplitude and corresponding shift of the resonant maxima to lower

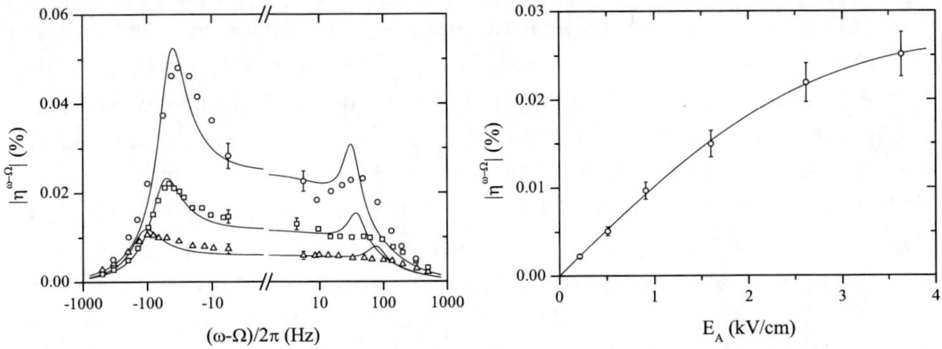

Figure 2.16. Frequency transfer function of the diffraction efficiency oscillations excited in $Bi_{12}SiO_{20}$ at the combination frequency $\omega - \Omega$ (left figure). The dependencies are obtained for three values of the applied dc field and for three corresponding frequencies of the applied ac field: $E_0 = 5.6$ kV/cm, $E_A = 1.4$ kV/cm, $\Omega/2\pi = 15$ kHz (△), $E_0 = 8.0$ kV/cm, $E_A = 2.6$ kV/cm, $\Omega/2\pi = 33$ kHz (□), and $E_0 = 16$ kV/cm, $E_A = 5.6$ kV/cm, $\Omega/2\pi = 130$ kHz (○). The approximation of the experimental dependencies by Eq. (2.60) is shown by the solid lines. Dependence of the diffraction efficiency amplitude $|\eta^{\omega-\Omega}|$ versus amplitude of an ac electric field E_A (right figure). $Bi_{12}SiO_{20}$, $E_0 = 8.0$ kV/cm, $\Omega/2\pi = 33$ kHz, $(\omega - \Omega)/2\pi = -54$ Hz. The solid line shows approximation by Eq. (2.61).

Table 2.2. Material parameters used for approximation of the experimental dependencies in Fig. 2.16 (left figure) by Eq. (2.60)

E_0, kV/cm	E_A, kV/cm	τ_M, ms	$\mu\tau$, cm^2/V
5.6	1.4	0.83	1.8×10^{-7}
8.0	2.6	1.7	1.3×10^{-7}
16	5.6	1.8	7.2×10^{-8}

frequencies. This behavior is the typical manifestation of the space charge waves [2]. The experimental curves are well fitted by Eq. (2.60) with the parameters presented in Table 2.2. Different values of the maxima in the region of negative and positive difference frequencies is an important peculiarity of the frequency transfer function (Fig. 2.16). One can notice another feature: the values of $\mu\tau$-product estimated from this frequency dependence noticeably decreases at large applied fields (Table 2.2). In fact, decrease of the quality factor of the space charge wave is observed. This decrease is most probably due to effects of trap saturation and it is not associated with any decay of $\mu\tau$-product. According to Ref. [75] trap saturation effects reduce quality factor at $E_0 > [(N_A/\epsilon\epsilon_0)(e/K^2\mu\tau + k_BT)]^{1/2}$. For $Bi_{12}SiO_{20}$ crystal with $N_A \simeq 10^{16}$ cm^{-3}, $\epsilon = 56$, and $\mu\tau \simeq 10^{-6}$ cm^2/V this reduction should appear at $E_0 \simeq 10$ kV/cm what well corresponds to the observed behavior. The proper description of the effect in trap saturation regime can be done by the appropriate modification of the basic equation set [Eqs. (2.1)-(2.5)] and subsequent calculations.

The dependence of the diffraction efficiency $|\eta^{\omega-\Omega}|$ measured at negative combination frequency versus amplitude of the applied ac field E_A is presented in Fig. 2.16 (right figure). This dependence seems to be linear in the region $E_A = 0 - 1.0$ kV/cm. The higher fields cause the noticeable saturation of the signal amplitude. This behavior can be explained as follows. The interaction of the ac field with the running photoconductivity grating is more efficient for higher amplitudes of an ac field, this also results in the growth of space charge wave. However, large ac fields perturb the propagation of the photoconductivity grating. As it is known, [38, 61] the excitation of running photoconductivity gratings occurs when the moving light pattern is synchronized with the propagation of photocarriers drifting in dc electric field. Higher values of the applied ac field comparable with the dc bias lead to considerable change of drift velocities, the synchronization is partially lost leading to the reduction of the running photoconductivity grating amplitude. This causes the saturation of the resulting space charge wave and corresponding detected signal. The obtained experimental dependence is approximated by Eq. (2.61) with $\mu\tau = 2.1 \times 10^{-7}$ cm^2/V.

Dependencies of the diffraction efficiency versus frequency of phase modulation $|\eta^{\omega-\Omega}(\omega)|$ were measured in $Bi_{12}SiO_{20}$ and $Bi_{12}TiO_{20}$ crystals for the constant combination frequency $\omega - \Omega \approx -\omega_{sc}$ (Fig. 2.17). The experimental curves demonstrate the behavior typical for running photoconductivity gratings: the growth of the dc electric field increases the amplitude of the detected signal and shifts the resonant peak to higher modulation frequencies (2.36). The resonant frequency values are slightly smaller than those obtained for the non-

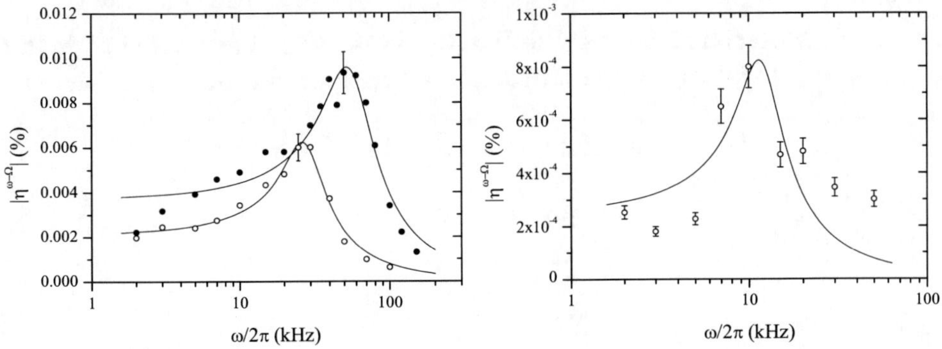

Figure 2.17. Dependencies of the diffraction efficiency amplitude $|\eta^{\omega-\Omega}|$ versus frequency of phase modulation ω measured in $Bi_{12}SiO_{20}$ (left figure) and $Bi_{12}TiO_{20}$ (right figure). $E_A = 0.92$ kV/cm. The measurements in $Bi_{12}SiO_{20}$ are carried out for two values of dc field and corresponding combination frequencies: $E_0 = 8.0$ kV/cm, $(\omega - \Omega)/2\pi = -48$ Hz (○) and $E_0 = 12$ kV/cm, $(\omega - \Omega)/2\pi = -38$ Hz (●). The measurements in $Bi_{12}TiO_{20}$ are carried out for $E_0 = 12$ kV/cm, $(\omega - \Omega)/2\pi = -10$ Hz. The approximations by Eq. (2.63) are shown by the solid lines.

steady-state photo-EMF experiments (Fig. 2.15). The approximation of the experimental dependencies by Eq. (2.63) allowed determination of the drift mobility of electrons as well as $\mu\tau$-product for $Bi_{12}SiO_{20}$ crystal (Table 2.3). The mobility measurements were also carried out in photorefractive $Bi_{12}TiO_{20}$. This crystal has sufficiently lower photoconductivity and corresponding resonant frequency of space charge oscillations ω_{sc}. Thus the minimal possible combination frequency $(\omega - \Omega)/2\pi = -10$ Hz was settled close to the resonant frequency. The amplitude of the detected signal was significantly lower than the one in $Bi_{12}SiO_{20}$ crystal due to the same reason: the smaller conductivity grating excites weak space charge wave. The electron mobility and the $\mu\tau$-product were estimated from the frequency dependence (Fig. 2.17): $\mu = 3.1 \times 10^{-3}$ cm^2/V s, $\mu\tau = 1.4 \times 10^{-7}$ cm^2/V ($T = 293$ K).

The obtained drift mobility values in $Bi_{12}SiO_{20}$ and $Bi_{12}TiO_{20}$ crystals are lower by 2-3 orders of magnitude than the actual band [39] and Hall [76] mobilities. Such difference is usually attributed to the process of strong trapping of

Table 2.3. Material parameters of the $Bi_{12}SiO_{20}$ crystal estimated from the frequency dependence of the diffraction efficiency amplitude $|\eta^{\omega-\Omega}(\omega)|$

E_0, kV/cm	T, K	μ, cm^2/V s	$\mu\tau$, cm^2/V
8.0	296	1.1×10^{-2}	1.9×10^{-7}
12	298	1.4×10^{-2}	1.1×10^{-7}

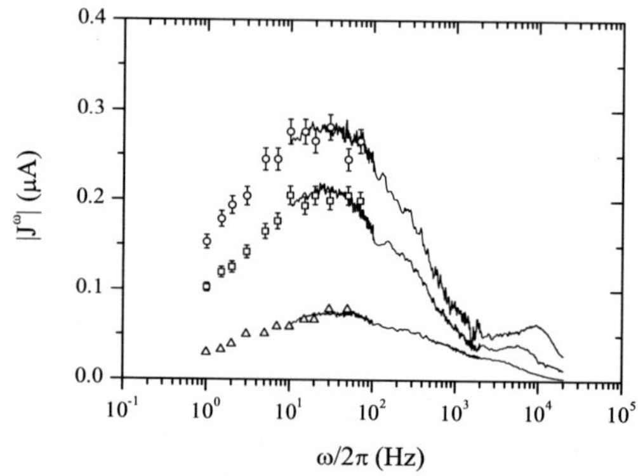

Figure 2.18. Frequency transfer functions of the non-steady-state photo-EMF excited in $Bi_{12}SiO_{20}$. Standard technique ($E_A = 0$). The dependencies are measured for three values of the applied dc electric field: $E_0 = 6$ kV/cm (△), 10 kV/cm (□), 14 kV/cm (○).

photoelectrons on the shallow levels [35] or large polarons influence [39].

In this section we also consider the application of another, namely, the non-steady-state photo-EMF technique for detection of the discussed processes in $Bi_{12}SiO_{20}$ at $\lambda = 442$ nm.

Figure 2.18 presents the frequency transfer functions of the non-steady-state photo-EMF measured in $Bi_{12}SiO_{20}$ crystal for three values of the applied dc field E_0. If the value of an applied electric field is increased the signal amplitude increases as well. Initially, the first resonant maximum appears in the low frequency region ($\omega/2\pi \sim 100$ Hz). When the dc field reaches $E_0 \simeq 6$ kV/cm the shoulder is observed on the frequency transfer function at $\omega/2\pi \sim 3$

kHz. The further increase of an applied field amplitude transforms this shoulder into the second resonant maximum. The growth of the applied field shifts the first and second maxima to lower and higher excitation frequency regions respectively. For example, the resonant frequencies equals $\omega_{sc}/2\pi \simeq 25$ Hz, $\omega_{pc}/2\pi \simeq 5.6$ kHz for the electric field $E_0 = 10$ kV/cm and $\omega_{sc}/2\pi \simeq 20$ Hz, $\omega_{pc}/2\pi \simeq 9.1$ kHz for the electric field $E_0 = 14$ kV/cm. It was shown [38] that such signal behavior is associated with the resonant excitation of the space charge and photoconductivity running gratings. The first resonance is very wide which can also be attributed to the nonlinear character of the space charge formation for large contrast of the interference pattern [74]. The light intensity inhomogeneity can be another reason for such behavior: strong light absorption ($\alpha \sim 30$ cm^{-1}) makes the space charge relaxation time in the surface layer smaller than in the crystal's bulk.

Direct measurements of the photocarrier's drift mobility were carried out using the non-steady-state photo-EMF technique. The mobility value was estimated from the position of the second resonant frequency for fixed K and E_0 values: $\mu \simeq 0.015$ cm^2/V s. This estimation is in good agreement with an electron mobility value $\mu = 0.016$ cm^2/V s measured earlier at the wavelength $\lambda = 458$ nm [38]. Nevertheless the large light absorption the parameters estimated from the high frequency resonant maximum seem to be rather correct since the resonant frequency and the shape of the resonant peak are not defined by parameters dependent on light intensity [Eqs. (2.57), (2.63)].

Let us proceed with the combined excitation of the space charge and photoconductivity gratings. The frequency of an external ac field was settled approximately equal to the frequency of the second resonant maximum: $\Omega/2\pi = 5.9$ kHz for $E_0 = 10$ kV/cm and $\Omega/2\pi = 8.8$ kHz for $E_0 = 14$ kV/cm. The amplitude of an external ac field ($E_A = 3.5$ kV/cm) was chosen to be considerably smaller than the dc field value.

Figure 2.19 (left figure) presents the frequency transfer function of the non-steady-state photo-EMF excited at difference combination frequency $\omega - \Omega$. Two maxima are clearly seen at this dependence: the first one is observed for the negative difference frequency $(\omega - \Omega)/2\pi = -(50 - 200)$ Hz ($E_0 = 10$ kV/cm), $(\omega - \Omega)/2\pi = -(20 - 200)$ Hz ($E_0 = 14$ kV/cm) and the second one – for the positive frequency $(\omega - \Omega)/2\pi = 50$ Hz ($E_0 = 10$ kV/cm), $(\omega - \Omega)/2\pi = 30$ Hz ($E_0 = 14$ kV/cm). The increase of the external dc

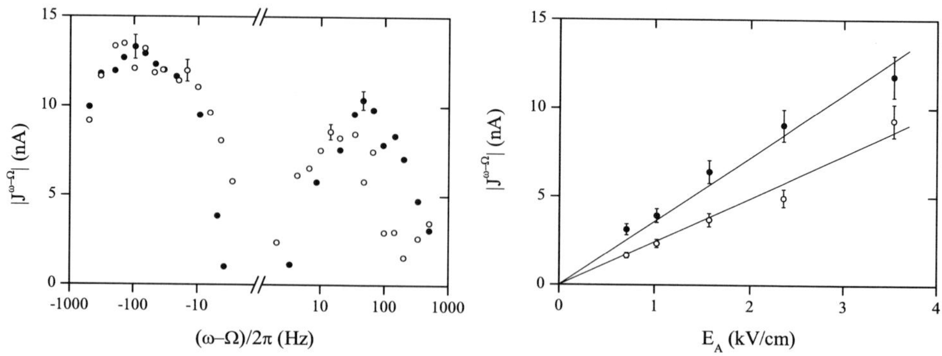

Figure 2.19. Frequency transfer function of the non-steady-state photo-EMF excited in $Bi_{12}SiO_{20}$ by the combination technique (left figure). The dependencies are obtained for two values of the applied dc field and for two corresponding frequencies of the applied ac field: $E_0 = 10$ kV/cm, $\Omega/2\pi = 5.9$ kHz (○) and $E_0 = 14$ kV/cm, $\Omega/2\pi = 8.8$ kHz (●). $E_A = 3.5$ kV/cm. Dependence of the non-steady-state photo-EMF amplitude $|J^{\omega-\Omega}|$ on the amplitude of the applied ac field (right figure). The measurements are carried out for two resonant frequencies: $(\omega - \Omega)/2\pi = -20$ Hz (●), $(\omega - \Omega)/2\pi = +20$ Hz (○). $E_0 = 14$ kV/cm, $\Omega/2\pi = 8.8$ kHz.

field causes the growth of the signal amplitude as well as the shift of the resonant maxima to lower frequencies. This behavior is typical for the space charge waves [2]. The resonant maxima turned out to be wide. This fact can be attributed to the nonlinear limitation of the space charge field for large contrasts of the interference pattern. The inhomogeneity of the light distribution through the sample thickness can be another reason of such behavior.

Figure 2.19 (right figure) presents the dependencies of the non-steady-state photo-EMF amplitude at resonant frequencies versus amplitude of the applied ac field. These dependencies seems to be linear in the investigated range of ac field amplitudes. The saturation of the signal amplitude can be expected for such values of an ac field E_A when $KL_A \sim 2$ [see Eqs. (2.55), (2.56)]. For $\mu\tau \sim 10^{-6}$ cm^2/V (see, for example Ref. [60]) and $K = 25$ mm^{-1} this saturation effect should be observed for $E_A \sim 8$ kV/cm.

In this section we have considered the combined excitation of two eigenmodes in semiconductor: the running gratings of the space charge and photoconductivity. The novel technique based on the diffraction efficiency measurements allowed us to detect the space charge wave originated from the nonlinear interaction of the running photoconductivity grating with the applied ac field.

The most important feature of this approach is the difference between the spatiotemporal characteristics of the arising space charge wave and the spatiotemporal characteristics of the external driving forces (illumination and applied field). In this sense, the experiments presented in the paper is another demonstration of the eigen-nature of space charge wave – the oscillation which properties are defined by the material parameters rather than external forces.

The combined excitation of the photoconductivity and space charge waves is very promising for the investigation of fast processes in high resistive semiconductors. The measurement of the actual drift mobility of photocarriers in widegap semiconductors with complicated structure of recombination levels is an important example of the tasks, where detection of high frequency signals is necessary. Indeed, the resonant frequency of the photoconductivity grating can reach 100 MHz in $Bi_{12}SiO_{20}$ crystal [61]. Due to the large values of Maxwell relaxation time ($\tau_M \sim 1$ ms) oscillations of the space charge field and corresponding diffraction efficiency should be negligible in this frequency range: the reduction coefficient is $(\omega\tau_M)^{-1}$. This makes observation of the high frequency resonances practically impossible. In contrast to the standard techniques the developed approach of combined excitation allows the detection of fast processes in the low-frequency region.

The nonlinear interaction of applied ac field with running gratings of the photoconductivity and space charge is not restricted to the results obtained in this paper. Further experimental investigations in other widegap semiconductors should be carried out, and more detailed theoretical analysis including trap saturation, optical activity and absorption effects should be performed to complete description of the phenomena discussed.

Chapter 3

Non-Steady-State Photo-EMF in Wide-Bandgap Semiconductors

1. SnS_2

Tin disulphide belongs to a class of compounds of the form MX_2, where M represents an element of group IV (Sn, Zr, Hf, Tl, etc.) and X an element of Group VIB (S, Se, Te) of the periodic table. These compounds crystallize in CdI_2, C-6 type crystal structure. The X atoms form densely packed hexagonal layers. The M atoms are arranged in layers with the same hexagonal structure to give a stacking sequence X-M-X, X-M-X, etc in the c-direction of the crystal. Each metal atom is surrounded by three X atoms of each layer so as to form octahedron. The bonds between these sandwich layers are weak and thought to be of Van der Waals type.

The optical properties of SnS_2 have been studied in Refs. [77–79]. The crystals of SnS_2 are highly sensitive in the visible region of spectrum, however, there are only several papers [80–82] dealing with the photoconductivity investigation in these crystals. Theoretical calculations of the energy band structure of SnS_2 crystals [83] showed that this material has several conductive sub-bands, that can have various energy with respect to the valence band (the energy interval depends on the crystal polytype). So the measurements at different light wavelengths should reveal this peculiarity. In this section we present the experimental results of the non-steady-state photocurrent measurements in SnS_2

crystals in the red (633 nm), blue (442 nm) and ultraviolet (325 nm) regions of spectrum [84, 85]. The material parameters such as type of photoconductivity, average diffusion length of photocarriers, dielectric relaxation time and the Debye screening length were determined using the non-steady-state photo-EMF technique [20, 42, 86–88]. Since this effect is based on the same principles as the dynamic holographic recording in photorefractive materials the results of this research can also be useful for the expansion of our understanding of the space charge field formation process in these media.

The observation of the non-steady-state photo-EMF in SnS_2 crystal was performed using the conventional experimental setup (Fig. 1.8, b). The detailed description of the experimental procedure is presented in Refs. [84, 85]. The SnS_2 crystal was grown by the vapor transport method using iodine as a transporting agent. The size of the sample was $6 \times 5 \times 0.042$ mm^3. The silver paste electrodes were painted on the front surface 6×5 mm^2. The distance between electrodes was about 1 mm. The band gap for SnS_2 is equal to 2.0-2.1 eV [78, 83, 89], the absorption coefficient for the illumination wavelengths is $\alpha \simeq 10$ cm^{-1} (for $\lambda = 633$ nm), $\alpha \sim 10^4$ cm^{-1} (for $\lambda = 442$, 325 nm) [78]. It means that the red light is absorbed slightly and practically uniformly within the crystal thickness whereas the near-surface absorption takes place for the blue and UV light. The optical axis of the crystal was perpendicular to the front surface so the grating vector laid in the plane where the crystal properties are uniform. The quoted value of the dielectric constant is $\epsilon = 6 - 20$, the electron mobility equals $\mu = 50$ cm^{2}V^{-1}s^{-1} [89]. We have measured the ϵ value for our crystal and it was found to be 7.0.

We observed the pronounced photocurrent signal J^{ω} in SnS_2 crystal for all illumination wavelengths. The signal to noise ratio (S/N) for the optimal spatial and temporal frequencies was about $S/N = 50$ dB for $\lambda = 633$ nm, $P_0 = 33$ mW, $\tau_{int} = 0.1$ s, $S/N = 43$ dB for $\lambda = 442$ nm, $P_0 = 0.28$ mW, $\tau_{int} = 0.3$ s, and $S/N = 40$ dB for $\lambda = 325$ nm, $P_0 = 3.8$ mW, $\tau_{int} = 1$ s (P_0 is the light power on the crystal surface, τ_{int} is the integration time).

It was shown in [20] that using non-steady-state photo-EMF technique one can define the type of photoconductivity. We compared the phase of the photocurrent signal in SnS_2 and $Bi_{12}SiO_{20}$ for all illumination wavelengths and spatial frequencies of the interference pattern $K = 0.1 - 20$ μm^{-1} and found out that SnS_2 has the same type of photoconductivity as $Bi_{12}SiO_{20}$, i.e., elec-

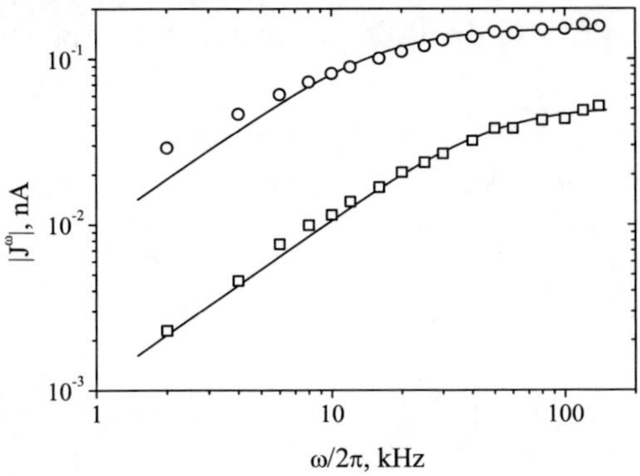

Figure 3.1. Frequency transfer functions of the non-steady-state photo-EMF in SnS_2 crystal. $\lambda = 633$ nm, $P_0 = 28$ mW, $K = 0.44\ \mu m^{-1}$ (○) and $\lambda = 442$ nm, $P_0 = 0.28$ mW, $K = 0.16\ \mu m^{-1}$ (□).

tron one [20]. Note that the sign of the dominant photocarriers can be obtained directly from the sign of the non-steady-state photocurrent signal, since J^ω has opposite signs for electrons and holes.

The experimental frequency dependencies of the photocurrent amplitude for wavelengths $\lambda = 633, 442$ nm are presented in Fig. 3.1. For low modulation frequencies the signal amplitude grows linearly with ω up to characteristic cut-off frequency ω_0; for higher frequencies of phase modulation the signal amplitude is frequency-independent. This behavior is typical for photorefractive materials and can be simply explained [20]. At high frequencies $\omega > \omega_0$ the photoconductivity distribution follows the movements of the interference pattern whereas the space charge field grating is nearly motionless. The condition for photocurrent excitation is optimal. For low excitation frequencies ($\omega < \omega_0$) the photocurrent decreases since both photoconductivity and electric field distributions can follow slow shifts of the interference pattern. The theory predicts the following frequency dependence of the photocurrent amplitude [20,42,86]:

$$J^\omega = \frac{Sm^2 \Delta\sigma_0 E_d}{2(1+K^2L_D^2)(1+K^2L_d^2)} \frac{-i\omega/\omega_0}{1+i\omega/\omega_0}, \tag{3.1}$$

where S is the sample cross-section, σ_0 is the average photoconductivity of the material, $E_d = Kk_BT/e$ is the diffusion field, k_B is the Boltzmann constant, T is the temperature, e is the electron charge, L_d is the diffusion length of photoelectrons and L_D is the Debye screening length. The cut-off frequency ω_0 is associated with the time of grating recording τ_{sc} as [42]:

$$\omega_0 = \frac{1}{\tau_{sc}} = \frac{1 + K^2L_D^2}{\tau_M(1 + K^2L_d^2)}, \tag{3.2}$$

here $\tau_M = \epsilon\epsilon_0/\sigma_0$ is the Maxwell relaxation time, ϵ_0 is the free space permittivity, ϵ is the static dielectric constant. According to (2) one can estimate the values of Maxwell relaxation time τ_M and photoconductivity σ_0 from the measurement of the cut-off frequency ω_0 for low spatial frequencies K ($K^2L_d^2$, $K^2L_D^2 \ll 1$). We performed such measurements and obtained for the cut-off frequency $\omega_0/2\pi$, Maxwell relaxation time τ_M, photoconductivity of the sample σ_0 and for average concentration of photocarriers n_0 the following: $\omega_0/2\pi = 16$ kHz, $\tau_M = 10$ μs, $\sigma_0 = 6.2 \times 10^{-6}$ Ω^{-1}m^{-1}, $n_0 = 7.8 \times 10^{15}$ m^{-3} ($\lambda = 633$ nm, $P_0 = 28$ mW) and $\omega_0/2\pi = 48$ kHz, $\tau_{SC} = 3.3$ μs ($\lambda = 442$ nm, $P_0 = 0.28$ mW, $K = 0.16$ μm^{-1}). The estimations were done for $\epsilon = 7.0$. The results of fitting are presented in the same figure. We observed linear dependence of the cut-off frequency ω_0 and photocurrent amplitude J^ω (for $\omega > \omega_0$) on the light power P_0 for the excitation wavelength $\lambda = 633$ nm and sub-linear behavior (ω_0, $J^\omega \propto P_0^{0.8}$) for $\lambda = 442$ nm. The photoconductivity σ_0 depends on the light intensity in the same way. We observed no reduction (for the modulation frequencies up to 150 kHz) of the photocurrent amplitude due to finiteness of the electron lifetime τ [42], so we can state that the lifetime of photocarriers τ is less than 1 μs.

We have performed the measurements of the non-steady-state photocurrent amplitude versus spatial frequency K (see Fig. 3.2) at low frequencies of phase modulation ($\omega \ll \omega_0$). According to Eq. (3.2) the cut-off frequency ω_0 is a decreasing function of K. Condition $\omega \ll \omega_0$ should be satisfied for whole range of spatial frequencies used. In our experiments we settled $K_{max} \simeq 20$ μm^{-1} and corresponding cut-off frequencies were $\omega_0/2\pi = 200$ Hz (for $\lambda = 633$ nm) and $\omega_0/2\pi = 2.8$ kHz (for $\lambda = 442$ nm). For both excitation wavelengths we fixed the frequency of phase modulation $\omega/2\pi = 30$ Hz so that condition $\omega \ll \omega_0$ was valid. For low frequency of phase modulation the photocurrent signal amplitude depends on spatial frequency of the interference pattern K

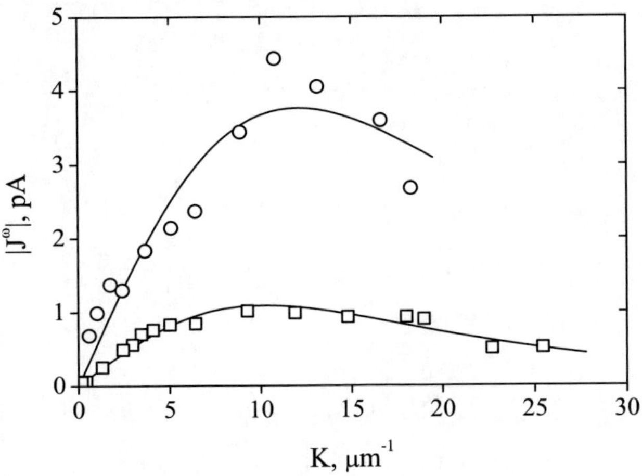

Figure 3.2. Dependencies of the non-steady-state photo-EMF on the spatial frequency measured for $\omega/2\pi = 30$ Hz $\ll \omega_0/2\pi$. $\lambda = 633$ nm, $P_0 = 34$ mW ($\circ$); $\lambda = 442$ nm, $P_0 = 0.42$ mW ($\Box$).

as [42]:

$$J^\omega \propto \frac{K}{(1 + K^2 L_D^2)^2}. \tag{3.3}$$

The signal amplitude J^ω peaks at spatial frequency $K = (\sqrt{3} L_D)^{-1}$. Approximation of the experimental results provides the following estimations for the Debye screening lengths: $L_D = 48$ nm (for $\lambda = 633$ nm) and $L_D = 61$ nm (for $\lambda = 442$ nm). Using the well-known expression for the Debye screening length [42] we calculated the density of capture centers: $N_A = 4.3 \times 10^{21}$ m^{-3} (for $\lambda = 633$ nm), $N_A = 2.6 \times 10^{21}$ m^{-3} (for $\lambda = 442$ nm).

The measurements of dependence $J^\omega(K)$ for $\omega > \omega_0$ were performed as well (Fig. 3.3). As it was shown in [20, 42] this dependence can also provide information about material parameters. The behavior of the nonstationary photocurrent is defined by the following expression [42]:

$$J^\omega \propto \frac{K}{(1 + K^2 L_D^2)(1 + K^2 L_d^2)}. \tag{3.4}$$

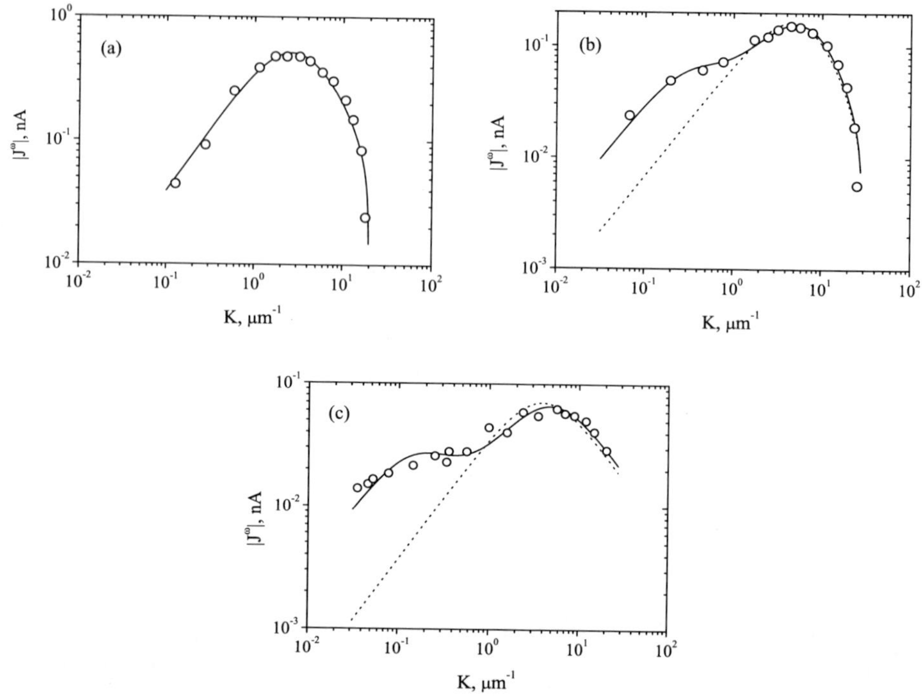

Figure 3.3. Dependencies of the non-steady-state photo-EMF on the spatial frequency measured for $\omega > \omega_0$. a – $\lambda = 633$ nm, $P_0 = 33$ mW, $\omega/2\pi = 100$ kHz; b – $\lambda = 442$ nm, $P_0 = 0.29$ mW, $\omega/2\pi = 140$ kHz; c – $\lambda = 325$ nm, $P_0 = 3.8$ mW, $\omega/2\pi = 94$ kHz. The dotted and solid lines show approximation by Eqs. (3.4) and (3.20), respectively.

The linear growth of the signal J^ω for low spatial frequencies K is due to increasing of the diffusion field E_d [20, 42]. Smoothing of the photoconductivity distribution at $K > L_d^{-1}$ results in decreasing of the non-steady-state photocurrent flowing through the crystal. For $L_d > L_D$ the dependence $J^\omega(K)$ has a maximum at $K \simeq L_d^{-1}$. Using equation (3.4) we determined the value of the diffusion length of photocarriers: $L_d = 0.38$ μm ($\lambda = 633$ nm).

As seen from Fig. 3.3, b the photocurrent excited at $\lambda = 442$ nm is not adequately described by Eq. (3.4) at low spatial frequencies (dash line). We

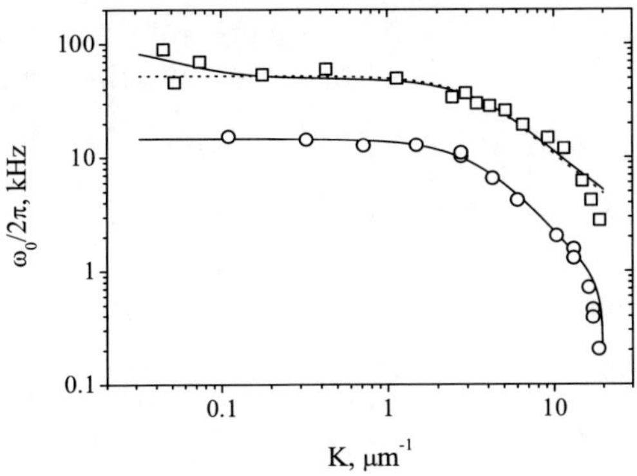

Figure 3.4. Dependence of the cut-off frequency ω_0 on the spatial frequency of the interference pattern. $\lambda = 633$ nm, $P_0 = 29$ mW (○); $\lambda = 442$ nm, $P_0 = 0.35$ mW (□).

tried to approximate the obtained results using the theory of the effect based on the concept of bipolar conductivity [86–88], but we did not obtain satisfactory correspondence as well. We think that the observed experimental dependencies can be explained by the presence of two types of photocarriers – light and heavy electrons. Equation (3.20) obtained below for such model of semiconductor describes experimental results at low spatial frequencies more correctly (see solid line in Fig. 3.3, b). Corresponding diffusion lengths of photocarriers and photoconductivity ratio were found: $L_{d1} = 0.15\ \mu$m, $L_{d2} = 2.0\ \mu$m, $\sigma_2/\sigma_1 = 4.3$.

Even more noticeable difference ($\sim$ 18 dB) between the experimental results and approximation by standard expression (3.4) is observed for excitation wavelength $\lambda = 325$ nm (Fig. 3.3, c). Approximation by Eq. (3.20) seems more reliable and provides the following estimations of material parameters: $L_{d1} = 0.14\ \mu$m, $L_{d2} = 5.0\ \mu$m, $\sigma_2/\sigma_1 = 11$. Hypothetically there is a possibility for the excitation of three types of electrons and in this case we get the following parameters: $L_{d1} = 0.16\ \mu$m, $L_{d2} = 0.85\ \mu$m, $L_{d3} = 11\ \mu$m, $\sigma_2/\sigma_1 = 2.1$ and $\sigma_3/\sigma_1 = 20$.

The dependence of cut-off frequency ω_0 versus spatial frequency of the interference pattern can also be used for estimation of the Maxwell relaxation time

τ_M and diffusion length of photocarriers L_d. We measured this dependence (see Fig. 3.4) and approximated it using Eq. (3.2) (solid line for $\lambda = 633$ nm, dash line for $\lambda = 442$ nm) and Eq. (3.18) (solid line for $\lambda = 442$ nm). Values τ_M and L_d for $\lambda = 633$ nm were found: $\tau_M = 11$ μs, $L_d = 0.25$ μm ($P_0 = 29$ mW). For $\lambda = 442$ nm we got the following results: $\tau_M = 1.6$ μs, $L_{d1} = 0.22$ μm, $L_{d2} = 20$ μm, $\sigma_2/\sigma_1 \simeq 1$ ($P_0 = 0.35$ mW).

Using the obtained value of τ_M we also estimated the average photoconductivity and concentration of free electrons for $\lambda = 442$ nm: $\sigma_0 \simeq 3.9 \times 10^{-5}$ Ω^{-1}m^{-1}, $n_0 \simeq 4.9 \times 10^{16}$ m^{-3}. Since there is no enough information about mobilities for each type of electrons it should be reminded that by using terms "light" and "heavy" electrons we imply that these carriers have different diffusion lengths.

It is well known that there are several minima (valleys) in the conduction band of many semiconductors (e.g. GaAs). The electrons in these valleys are characterized with different effective masses, lifetimes and mobilities. In this section we analyze the effect of the non-steady-state photo-EMF for the model of semiconductor with two channels of charge migration characterized by different transport parameters.

Let us suppose that the crystal is illuminated by an oscillating interference pattern:

$$I(x,t) = I_0 \left[1 + \frac{m(t)}{2} \exp(iKx) + \frac{m^*(t)}{2} \exp(-iKx) \right], \quad (3.5)$$

where

$$m(t) = m \exp(i\Delta \cos \omega t) \quad (3.6)$$

is the contrast of the interference pattern.

Let us consider the model of semiconductor with two types of charge carriers (electrons) characterized with mobilities μ_1, μ_2 and lifetimes τ_1, τ_2. The excitation and recombination of electrons are assumed to be linear. The transitions of electrons between valleys are neglected. Then the balance equations for the corresponding concentrations of electrons n_1, n_2 are as follows:

$$\frac{\partial n_{1,2}}{\partial t} = g_{1,2} - \frac{n_{1,2}}{\tau_{1,2}} + \mu_{1,2} \frac{\partial}{\partial x}(n_{1,2}E) + D_{1,2} \frac{\partial^2 n_{1,2}}{\partial x^2}, \quad (3.7)$$

where g_1, g_2 are the generation rates of electrons and $D_{1,2} = (k_B T/e)\mu_{1,2}$ are their diffusion coefficients.

In order to calculate the distributions of the carrier density and space charge field the continuity and Poisson equations should be added [2]. Differentiating the Poisson equation with respect to time and using the continuity equation we get

$$\epsilon\epsilon_0 \frac{\partial^2 E}{\partial x \partial t} = -\frac{\partial j_e}{\partial x}. \tag{3.8}$$

The current density j_e created by the transport of electrons of both types equals

$$j_e = e\mu_1 n_1 E + eD_1 \frac{\partial n_1}{\partial x} + e\mu_2 n_2 E + eD_2 \frac{\partial n_2}{\partial x}. \tag{3.9}$$

The calculations of the photocurrent and electric field are performed using the technique proposed in [2, 19]. For low contrast of the interference pattern $m \ll 1$ values $n(x,t)$ and $E(x,t)$ can be written using their complex amplitudes [2]:

$$n_{1,2}(x,t) = n_{1,2}^0 \left[1 + \frac{a_{1,2}(t)}{2} \exp(iKx) + \frac{a_{1,2}^*(t)}{2} \exp(-iKx)\right], \tag{3.10}$$

$$E(x,t) = \frac{E_{sc}(t)}{2} \exp(iKx) + \frac{E_{sc}^*(t)}{2} \exp(-iKx). \tag{3.11}$$

Here $n_{1,2}^0 = g_{1,2}^0 \tau_{1,2}$ are the average concentrations of electrons of both types. Then the nonstationary photocurrent equals

$$j(t) = \sum_{q=1,2} e\mu_q n_q^0 [a_q(t) E_{sc}^*(t) + a_q^*(t) E_{sc}(t)]. \tag{3.12}$$

Substituting Eqs. (3.10), (3.11) into Eqs. (3.7)-(3.9) we can calculate complex amplitudes $a_{1,2}(t)$ and $E_{sc}(t)$. For the simplicity let us suppose the electron densities are quasi-stationary: $\partial n_{1,2}/\partial t \simeq 0$ [2]. Then complex amplitudes $a_{1,2}(t)$ are easily obtained [2, 19]:

$$a_{1,2}(t) = \frac{m(t) + iK\mu_{1,2}\tau_{1,2}E_{sc}}{1 + K^2 L_{d1,d2}^2}. \tag{3.13}$$

Here $L_{d1,d2} = (D_{1,2}\tau_{1,2})^{1/2}$ are the diffusion lengths of electrons of both types. Further substitution of Eqs. (3.10), (3.11) into(3.8), (3.9) with (3.13) leads to the equation describing the process of space charge grating formation:

$$\frac{dE_{sc}}{dt} = -\left[\frac{\sigma_1}{1 + K^2 L_{d1}^2} + \frac{\sigma_2}{1 + K^2 L_{d2}^2}\right] \frac{imE_d + E_{sc}}{\epsilon\epsilon_0}. \tag{3.14}$$

Here $\sigma_{1,2} = e\mu_{1,2}n^0_{1,2}$ are the average values of the conductivity corresponding to the two types of electrons. The obtained equation differs from the classic equation of hologram recording [2] by the presence of two components in the brackets.

Let us suppose the amplitude of phase modulation to be small: $\Delta \ll 1$, then $m(t) \simeq m[1 + i0.5\Delta \exp(i\omega t) + i0.5\Delta \exp(-i\omega t)]$. Amplitudes $a_{1,2}(t)$ and $E_{sc}(t)$ can be presented as

$$a_{1,2}(t) = a^0_{1,2} + a^+_{1,2} \exp(i\omega t) + a^-_{1,2} \exp(-i\omega t), \quad (3.15)$$

$$E_{sc}(t) = E^0_{sc} + E^+_{sc} \exp(i\omega t) + E^-_{sc} \exp(-i\omega t), \quad (3.16)$$

Substituting Eq. (3.16) into Eq. (3.14) we obtain

$$E^0_{sc} = -imE_d, \qquad E^{\pm}_{sc} = \frac{m\Delta}{2} \frac{E_d}{1 \pm i\omega/\omega_0}. \quad (3.17)$$

Here we introduced the cut-off frequency of the non-steady-state photo-EMF:

$$\omega_0 = \frac{1}{\epsilon\epsilon_0} \left[\frac{\sigma_1}{1 + K^2 L^2_{d1}} + \frac{\sigma_2}{1 + K^2 L^2_{d2}} \right]. \quad (3.18)$$

It is seen from the expression for E^0_{sc} that amplitude of the stationary space charge field grating recorded in the semiconductor with two types of carriers (with the same sign) is exactly equal to the standard value $-imE_d$ [2]. It is worth to point out that the amplitude of the grating recorded in semiconductor with bipolar conductivity decreases due to partial compensation of the space charges created by electrons and holes [2]. After substitution of Eq. (3.17) into Eq. (3.13) one can obtain expressions for the complex amplitudes of the electron density:

$$a^0_{1,2} = m, \qquad a^{\pm}_{1,2} = \frac{im\Delta/2}{1 \pm i\omega/\omega_0} \left[1 \pm \frac{i\omega/\omega_0}{1 + K^2 L^2_{d1,d2}} \right]. \quad (3.19)$$

At last using the obtained expressions for $a_{1,2}$ and E_{sc} we can calculate the amplitude of the non-steady-state photo-EMF:

$$j^{\omega} = \frac{m^2 \Delta E_d}{2} \frac{-i\omega/\omega_0}{1 + i\omega/\omega_0} \left[\frac{\sigma_1}{1 + K^2 L^2_{d1}} + \frac{\sigma_2}{1 + K^2 L^2_{d2}} \right]. \quad (3.20)$$

Let us specify the main features of the non-steady-state photo-EMF in semiconductors with two types of charge carriers. The frequency transfer function is standard, i.e., there are linear growth of the signal at low frequencies of phase modulation ($\omega \ll \omega_0$) and frequency-independent region at higher frequencies ($\omega \gg \omega_0$). The dependence of the photocurrent on the spatial frequency (for $\omega \gg \omega_0$) can have different shapes for various relations between parameters of the carriers (L_{d1}, L_{d2}, σ_1, σ_2). We just point out that for low spatial frequencies ($K^2 L^2_{d1,d2} \ll 1$) the amplitude of the photocurrent j^ω ($\omega \gg \omega_0$) and cut-off frequency ω_0 are defined by electron type producing the dominant contribution in the total photoconductivity, whereas for high spatial frequencies ($K^2 L^2_{d1,d2} \gg 1$) it is defined by electron type with higher generation rate (compare with Ref. [86]). In contrast to the case of bipolar conductivity [86], the photo-EMF signal does not change its sign for any spatial frequency.

It is important to point out that crystal SnS_2 has a high photosensitivity: corresponding cut-off frequencies are in kilohertz range for all illumination wavelengths and powers used. For the well-known photorefractive sillenite-type crystals $Bi_{12}Si(Ge,Ti)O_{20}$ this value is in the range of hundreds of hertz for the same light intensities.

The Debye screening length value is rather small so the space charge field formation without saturation is possible in wide range of spatial frequencies $K = 0 - 10\ \mu m^{-1}$. In spite of the high photoconductivity this material has small diffusion length, especially for $\lambda = 442$ nm (for comparison: typical diffusion length of photoelectrons in photorefractive $Bi_{12}SiO_{20}$ at $\lambda = 442$ nm is about 2 μm).

The non-steady-state photo-EMF excited in SnS_2 at $\lambda = 633$ nm is well described in the frames of the standard model of semiconductor. Indeed, such illumination generates conduction band electrons from the impurity centers since the energy of photons ($h\nu = 1.96$ eV) is not enough for the interband transitions (forbidden band is of $2.07 - 2.18$ eV).

For the light with $\lambda = 442$ nm the energy of photons is of 2.81 eV. Such energy is enough for the generation of electrons not only to the minimum with lowest energy (it corresponds to the indirect transition in SnS_2) but also to the minimum which energy is higher by ~ 0.8 eV (direct transitions [78, 90]). Although there in no data concerning effective masses or mobilities of electrons in these valleys, we can suppose at least that diffusion lengths of electrons are different, so the utilization of the considered model is reasonable.

The dependence of the photocurrent on the spatial frequency measured at shorter wavelength $\lambda = 325$ nm ($h\nu = 3.82$ eV) differs from the standard one even more noticeably. In this case the carriers in layers with other polytype modification can also contribute to the total photocurrent. For example, layers of polytype $9R$ with forbidden band of 3.38 eV [91] can be present between layers of basic polytype modification $1H$.

The assumption of independent recombination and transport processes used in the theoretical analysis implies that the inter-sub-band relaxation time τ_{21} should exceed the electron lifetimes for the both sub-bands, i.e., $\tau_{21} \gg \tau_1, \tau_2$. We think that such relation is possible because the minimums of these sub-bands are shifted in the k-space [83] and the presence of the phonon is required for the inter-sub-band transition.

2. BN

Let us present experimental investigations of the non-steady-state photocurrents excited in pyrolytic crystal of boron nitride at the wavelength $\lambda = 532$ nm. The experiments are carried out for the diffusion (zero electric field) and drift mechanisms of recording (in an external ac electric field). The average photoconductivity and diffusion length of photoholes are estimated from the frequency transfer functions and dependence of the signal on electric field amplitude.

Modern optoelectronics requires new materials to manufacture devices operating at different wavelengths, electric fields, temperatures, etc. Wide-gap semiconductors such as GaN, ZnSe with band gap $E_g > 2.5$ eV are of particular importance for applications in semiconductor lasers and light emitting diodes for blue and ultraviolet spectral regions. The same materials are to be utilized for production of photoresistors and photodiodes. Boron nitride crystals are characterized by an unique combination of physicochemical properties that advance its wide application in various areas of science and technology. Due to its high thermal stability boron nitride is a widely used material in vacuum technology. It has been employed in microelectronic devices, for x-ray lithography masks, and as a wear-resistant lubrificant. The hexagonal phase is also the underlying structure of BN nanotubes, which are systems of growing interest at present. [92]

The chemical vapor deposition method [93] provides bulk samples of anisotropic modifications of boron nitride with high order degree of crystal lattice and nanostructure. Polarization infrared spectroscopy of well-ordered samples shows the existence of interlayer ion-covalent bonds which define the one-dimensional order-disorder phase transition between hexagonal and rhombohedral modifications of BN. The most perfect samples of pyrolytic BN crystallize in the form of large monocrystal layers of rhombohedral modification with transverse dimensions of 1 cm^2, thickness of 2000 Å and translation period $C_R = 3d$, where d is an interlayer distance. The crystal axes of the adjacent crystallites are correlated and the deviation of their c-axes does not exceed 1°. The samples also contain about 20% of hexagonal phase with unordered interlayer bonds with the typical translation period of $C_H \simeq 2d$. The dimensions of hexagonal crystallites are $\sim$ 100 Å with the off-orientation degree of their hexagonal axes 15°. Typical microstructure of an as grown pyrolytic BN is presented in Fig. 3.5.

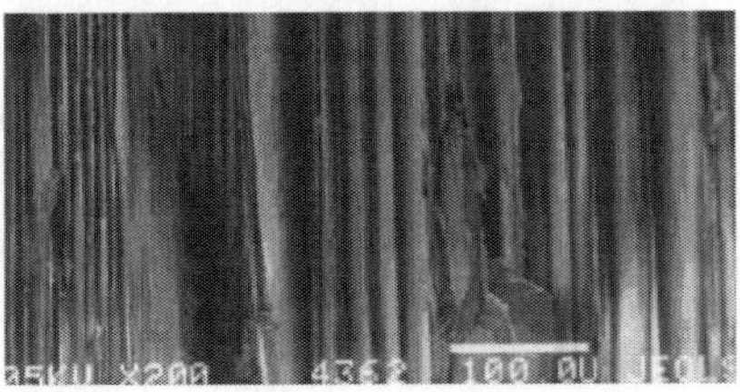

Figure 3.5. The microstructure of the investigated pyrolytic boron nitride. The c-axis is in the plane of the figure (parallel to the scale line of 1 μm).

Optical and electrical properties of boron nitride were investigated in a number of papers (see, for example, Refs. [94–99]). However, this does not mean that layered BN is a well investigated material: the scattering of the experimental data is significant. For example the estimated values of the band gap vary in the range 3.6-5.9 eV. [95] The absorption spectrum *of an as grown* rhombohedral BN measured in wide energy range corresponds to direct transitions and provides estimation of the forbidden gap to be $E_g = 5.67$ eV.

The measurements of the nonstationary holographic currents in BN crystal were carried out using standard experimental setup (Fig. 1.8, b). The detailed description of the arrangements is presented in Ref. [100].

First let us consider the diffusion regime of photocurrent excitation when the external electric field is not applied to the sample. The sign of the detected signal in the investigated frequency range ($\omega/2\pi = 1.5 - 200$ Hz) corresponds to the hole photoconductivity of the crystal.

The frequency transfer functions of the nonstationary photocurrent are shown in Fig. 3.6. These dependencies demonstrate typical behavior of the signal excitation, i.e. the linear growth for low modulation frequencies and the plateau for higher frequencies. Such behavior is an important manifestation of adaptive nature of space charge formation in photoconductive materials. As seen from Fig. 3.6 experimental points are in satisfactory agreement with the theoretical dependencies calculated using following expression [20]:

$$J^{\omega} = \frac{Sm^2\Delta\sigma_0 E_D}{2(1+K^2L_D^2)} \frac{-i\omega/\omega_0}{1+i\omega/\omega_0}, \tag{3.21}$$

where

$$\omega_0 = [\tau_M(1+K^2L_D^2)]^{-1} \tag{3.22}$$

is the cut-off frequency.

For high light intensities we observed the growth of the photocurrent amplitude and corresponding shift of the cut-off frequency (Fig. 3.7), what is associated with the increase of the crystal's photoconductivity. These dependencies are slightly nonlinear in the investigated range of light intensities: $|J^{\omega}| \propto I_0^{0.75}$ and $\omega_0 \propto I_0^{0.64}$. The non-steady-state photo-EMF technique provides the possibility for determination of crystal's photoconductivity from the position of corresponding cut-off frequency ω_0 [Eq. (3.22)] measured at low spatial frequency $KL_D \ll 1$. For the layered BN crystals the dielectric constant equals $\epsilon_{\perp} = 6.85 - 7.04$ (Ref. [94]), the diffusion length is measured below ($L_D = 35 - 50$ nm, $KL_D = 0.03 - 0.05$) so we obtain for the photoconductivity value of $\sigma_0 = (0.9 - 2.4) \times 10^{-10}\ \Omega^{-1}\text{cm}^{-1}$ ($I_0 = 0.22 - 0.86$ W/cm^2). This estimate is about two orders of magnitude lower than the one for $Bi_{12}SiO_{20}$ crystal which is a model object for non-steady-state photo-EMF technique [20]. We should note here that Eqs. (3.21), (3.22) were obtained for case when the crystal is illuminated uniformly. In our experiments the light penetration into the crystal volume is prohibited by the layered structure of the material which operates like a multilayer dielectric mirror (light penetration depth $\sim$1 μm). The

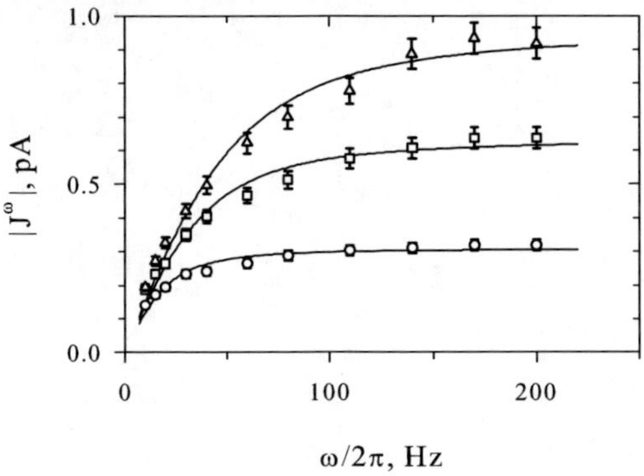

Figure 3.6. Frequency transfer functions of the nonstationary photocurrent measured for different intensity levels of the incident light: $I_0 = 0.22$ W/cm^2 (○), 0.44 W/cm^2 (□), 0.86 W/cm^2 (△). BN, $\lambda = 532$ nm, $K = 0.90\ \mu$m^{-1}, $E_0 = 0$. Solid lines show the approximation by Eq. (3.21).

influence of the layer geometry on the holographic photocurrent excitation can be noticeable only for low spatial frequencies of the interference pattern and reveals as an additional coefficient $[1 + (\epsilon K d/2)^{-1}]$ after τ_M in the expression for cut-off frequency (d is the thickness of the layer where photocurrent is excited) [101]. Our estimations shows that this coefficient should not exceed 1.3 for $K \geq 0.90\ \mu$m^{-1}. Another peculiarity of the photocurrent excitation in layered crystals can reveals as follows: the crystal is similar to a multilayered dielectric mirror which reflectance and transmittance are sensitive to the wavelength and incidence angle. So the dependencies of the nonstationary photocurrent and cut-off frequency versus spatial frequency (incidence angle) can differ from the theoretical ones. Although such effects are smoothed away in crystals with varying thickness of layers an additional experimental error can occur.

The dependence of the cut-off frequency versus spatial frequency of the interference pattern is presented in Fig. 3.8. As follows from Eq. (3.22) such a dependence has a breakpoint at spatial frequency corresponding to the inverse value of diffusion length. We have approximated the experimental dependence by Eq. (3.22) and found the diffusion length of holes to be $L_D \simeq 35$ nm. The

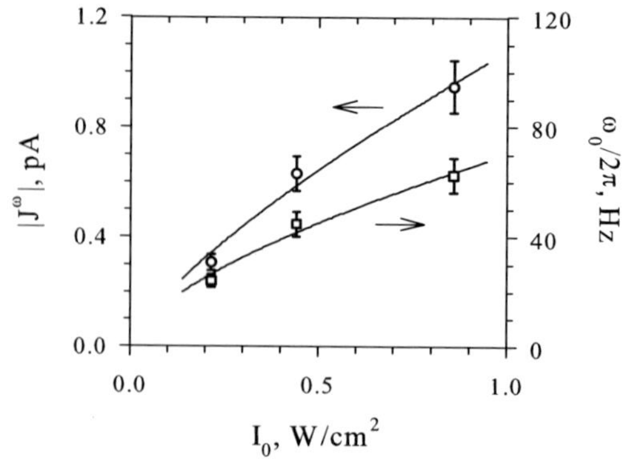

Figure 3.7. Dependencies of the signal amplitude (○) and cut-off frequency (□) on the light intensity. BN, $\lambda = 532$ nm, $K = 0.90\ \mu\text{m}^{-1}$, $\omega/2\pi = 200$ Hz, $E_0 = 0$. Solid lines show the approximation by power functions: $J^\omega \propto I_0^{0.75}$, $\omega_0 \propto I_0^{0.64}$.

fast decay of the cut-off frequency at high K is most probably resulted from the light reflectance increasing at large incident angles. The slight decrease of light intensity and corresponding cut-off frequency is typical for low spatial frequencies as well: when the crystal is moved far from the mirrors the light divergence become noticeable. So the estimation of the diffusion length obtained by this method is rather rough and can be considered as the lower bound for this parameter.

A new approach for investigation of materials with low conductivity and for determination of photocarrier transport parameters was developed in Ref. [60]. It is based on the excitation of the non-steady-state photocurrents in an external alternating electric field. This novel technique allow us to avoid variations of the spatial frequency in wide range accompanied with the time consuming readjustments of optical set-up and hardly controlled experimental errors. The dependence of the holographic photocurrent on the effective value of the external field is presented in Fig. 3.9. This dependence can be approximated by the following expression obtained for $\omega > \omega_0$ [60]:

$$J^\omega(E_0) = -\frac{Sm^2\Delta\sigma_0 E_D}{2(1+K^2L_D^2)}\frac{1+E_0^2/(E_D^2+E_L^2)}{1+E_0^2/(E_D+E_M)^2}, \tag{3.23}$$

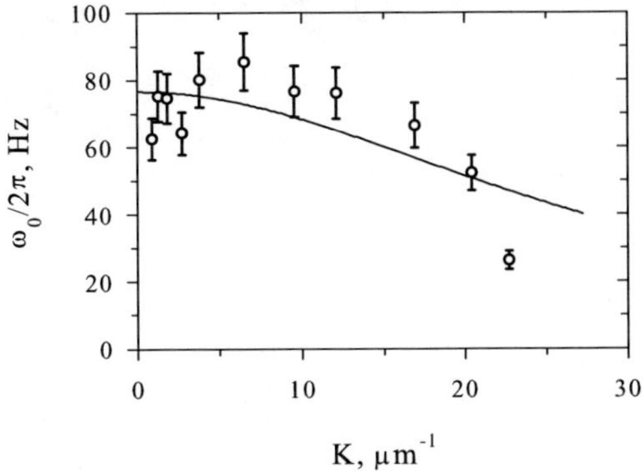

Figure 3.8. Dependence of the cut-off frequency on the spatial frequency of the interference pattern. BN, $\lambda = 532$ nm, $I_0 = 0.86$ W/cm^2, $E_0 = 0$. Solid line shows the approximation by Eq. (3.22).

here $E_L = (k_B T/e\mu\tau)^{1/2}$, $E_M = (K\mu\tau)^{-1}$ are the characteristic values of the electric field (μ is the mobility and τ is the lifetime of photocarriers). For the investigated BN crystal they equal $E_L = 5.0$ kV/cm ($\mu\tau = 1.0 \times 10^{-9}$ cm^2/V, $L_D = 50$ nm), $E_M = 5.2$ kV/cm. Characteristic fields E_L, E_M are not independent parameters by definition. In practice, however, it is reasonable to consider them separately since saturation effects can influence the position of the second breakpoint on dependence $|J^\omega(E_0)|$ and add error in measurement of E_M.

We have also investigated the influence of a dc electric field on non-steady-state photocurrent amplitude in BN crystal. No considerable effect was observed up to the breakdown values of electric field amplitudes (~ 15 kV/cm). The similar difference between application of ac and dc external fields is reported above where ac electric field contributed to resonant excitation of photoconductivity gratings while dc one did not. Several reasons of such behavior can be pointed out: first there are voltage drop on the blocking contact and inhomogeneity of electric field distribution in the crystal placed in dc electric field. If ac field has rather high frequency ($\Omega > \tau_M^{-1}$) these effects are imperceptible. This difference can be associated with the complicated processes of conductivity relaxation. The photoconductivity response to the amplitude mod-

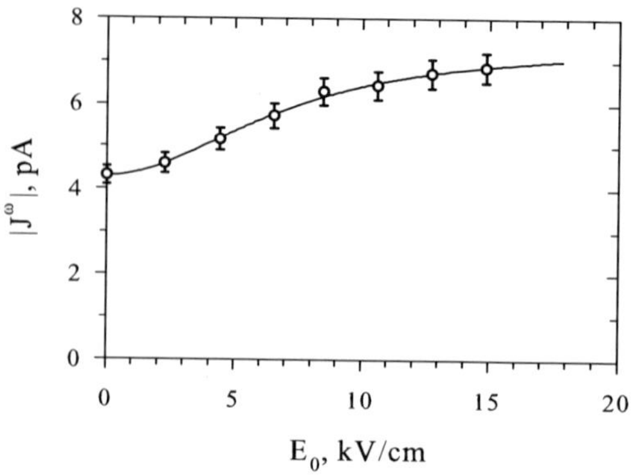

Figure 3.9. Dependence of the photocurrent amplitude on the effective value of applied ac electric field. BN, $\lambda = 532$ nm, $K = 6.9\ \mu\text{m}^{-1}$, $I_0 = 1.1$ W/cm^2, $\omega/2\pi = 200$ Hz. Solid line shows the approximation by Eq. (3.23).

ulated light is shown in Fig. 3.10. As seen there are two frequency independent regions $\omega/2\pi = 1.5 - 7$ Hz and $\omega/2\pi = 150 - 3000$ Hz. The frequency of external field $\Omega/2\pi = 560$ Hz corresponds to the second region so the effective photoelectric parameters governing the charge transport may not coincide for the cases of ac and dc fields applied. It should be noted that the origin of the extrinsic hole photoconductivity for $\lambda = 532$ nm is unknown. Perhaps it can be associated with carbon impurities which serve as luminescent centers and create local levels located $1.2 - 1.6$ eV above the valence band [102] or with nitrogen vacancies creating several donor and acceptor levels [103].

Besides the nonresonant excitation of photocurrent presented in this paper the resonant technique exists (see, for example, Ref. [20]). The latter approach provides the possibility of direct determination of photocarrier's mobility. The resonant excitation of the non-steady-state photocurrent is accompanied by a significant growth of the photocurrent amplitude and can be observed in the long drift length regime when the product $KL_0 \gg 1$ (here L_0 is the drift length of photocarriers). The application of an external electric field did not significantly increase the photocurrent amplitude in BN as was in our previous experiments with $Bi_{12}SiO_{20}$ crystals. The influence of the dc field was not observed at all. This fact can be attributed to the small value of the $\mu\tau$-product of BN. The

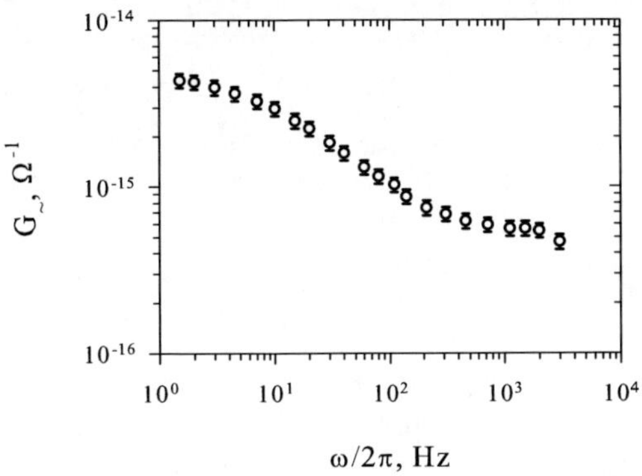

Figure 3.10. Frequency transfer function of the photoconductivity response. BN, $\lambda = 532$ nm, $I_0 = 0.32$ W/cm^2.

measured transport parameters ($L_D = 35 - 50$ nm) allow estimation of the characteristic value of an ac or dc field amplitudes providing resonant excitation of the non-steady-state photocurrent ($KL_0 = 10$, $K = L_D^{-1}$): $E_0 = 50 - 70$ kV/cm. This value is considerably larger than the breakdown field of the investigated BN sample (~ 15 kV/cm). The resonant excitation technique can be used for investigation of BN crystals in future when bulk single crystals of boron nitride will be available.

To summarize, we have studied the non-steady-state photocurrent generation in new widegap crystal of boron nitride. The photoconductivity and diffusion length of photoholes were estimated: $\sigma_0 = (0.9 - 2.4) \times 10^{-10}\ \Omega^{-1}\text{cm}^{-1}$ ($I_0 = 0.22 - 0.86$ W/cm^2), $L_D = 35 - 50$ nm [100].

3. GaN within Porous Glass

The utilization of noncrystalline materials instead of single crystal analogs is the new trend in modern optics and optoelectronics. Such materials are practically feasible and cheap, they can be chemically stable, and their photoelectric properties can be easily varied.

The porous media can be considered as new objects attracting great interest of the specialists in the solid state physics, optics, and semiconductors. Asbestoses, ceramics, and porous glasses are typical examples of this class of materials. The nanopores in these materials with the average diameter ranging from 10 to 100 nm may be doped with almost any gas, liquid or a liquid solution of a solid. The physical properties of fabricated structures can be different. As the result the material may be insulating or conductive, transparent, absorbing or reflecting. The nanoscale of the filling component advances observation of such effects as charge localization, quantum size effects, Peierls phase transition.

There are applications and devices where semiconductors with low conductivity are required (wafers for high-power microwave transistors, X-ray detection, holographic recording including optical data storage and processing, nondestructive testing). Gallium nitride (GaN) is considered as one of the most perspective material for such areas. Undoped GaN samples are characterized with rather high conductivity $\sigma = 1.2 \times 10^1 - 3.2 \times 10^2\ \Omega^{-1}\text{cm}^{-1}$ [104, 105]. A certain success has been achieved in fabrication of insulating GaN films using hydride vapor phase epitaxy (HVPE) methods. The conductivity of that samples reaches $\sigma \sim 10^{-12}\ \Omega^{-1}\text{cm}^{-1}$ [106]. The production of a composite material filled with nanoparticles of GaN can be considered as another way to obtain insulating samples (especially bulk ones).

The application of the standard methods for semiconductor characterization (see, for example, Ref. [44]) is often restricted by the space charge formation near the blocking contacts. This problem is of vital importance for investigation of high-resistive materials where space charge formation time can be very large and introduce inadmissible errors. The novel techniques based on holographic recording are more suitable for semi-insulating samples with low conductivity and mobility of photocarriers [2]. The non-steady-state photoelectromotive force (photo-EMF) technique based on the detection of the alternating electric current arising in the sample illuminated by an oscillating interference pattern is the most versatile and advanced method [20]. Since the current is resulted from the periodic relative shifts of the photoconductivity and space charge gratings the technique based on this effect allows determination of the number of photoelectric parameters (conductivity, carrier sign, lifetime, diffusion length, and drift mobility) and can be applied for the investigation of both non-centrosymmetric and centrosymmetric media.

The investigation of the new class of the semi-insulating materials, namely, porous glasses filled with nanoscaled components using the non-steady-state photo-EMF technique is the main goal of this section.

We used the conventional experimental arrangement for investigations of the non-steady-state photo-EMF (Fig. 1.8, b). One can find details of the setup in Refs. [107, 108]. A porous silica glass is a random structure of two interpenetrating phases, namely the solid and pore networks [109]. The pores in the glasses are connected to each other and the pore size distribution is narrow. The structure is obtained due to spinodal decomposition of the two phases SiO_2 and B_2O_3+Na_2O. After the glass is heat-treated and annealed, the B_2O_3-rich phase is removed by leaching with acid, leaving an almost pure SiO_2 skeleton. The pore sizes were tested by the mercury intrusion porosimetry and the average pore diameter was found to be 7 ± 1 nm. Gallium nitride has been synthesized directly inside the pores by chemical bath deposition technique similar to the described elsewhere [110]. At first, solid precursor (Ga_2O_3) was embedded into the pores of vycor template. We used alcohol solutions of gallium salts (gallium nitrate or gallium formate) as a soluble precursors to infiltrate Ga_2O_3. The filling procedure was repeated periodically. To produce GaN the vycor sample filled with gallium oxide was annealed in presence of ammonia (1000 Torr) at 750 °C for 50 h. The fill factor of pores with GaN was measured by gravimetric method and has been found to be 25%.

The structural state of the composite was determined by X-ray diffraction (XRD) measurements. The XRD patterns were obtained by using CuK_α radiation (Ni filter). Polycrystalline germanium was used as a standard. Figure 3.11 (curve 3) shows XRD patterns of the nanocomposite. As evident from these angular dependencies of the intensities scattered the material synthesized in the pores is the hexagonal GaN. Small amounts of crystalline phases such as α-Ga_2O_3 (JCPDS 6-0503), β-Ga_2O_3 (JCPDS 11-370), and α-tridymite (JCPDS 18-1170) also present in the composite. The analysis of Fig. 3.11 has also provided the typical size of GaN particles: 15 ± 5 nm. Note that this value is about 2 times larger than average pore dimension. Similar feature was observed earlier in the samples of Pb within porous glass matrix [111] and it was attributed to the elongated form of the nanoparticles.

The presence of the effect of the non-steady-state photo-EMF in new class of materials – porous glasses – is the first result that should be pointed out [107, 108]. In spite of the small signal amplitudes we can obtain reliable results by the use of high sensitive lock-in nanovoltmeter with integration time up to

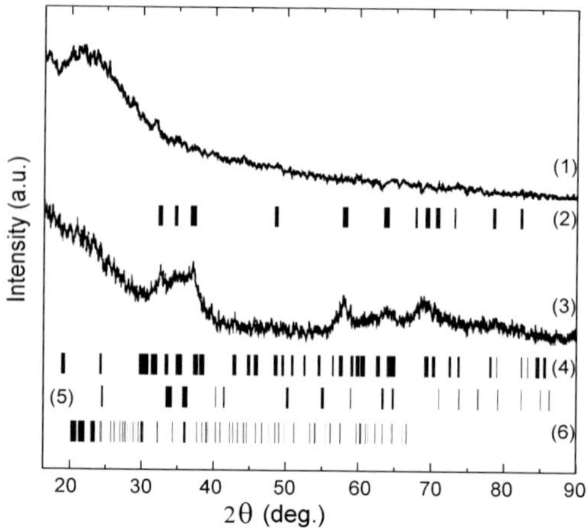

Figure 3.11. X-ray diffraction patterns. 1 – bare porous glass, 2 – bulk hexagonal GaN (JCPDS 2-1078), 3 – nanocomposite porous glass-GaN, 4 – α-Ga_2O_3 (JCPDS 6-0503), 5 – β-Ga_2O_3(JCPDS 11-370), 6 – α-tridymite (JCPDS 18-1170).

100 s. We have determined the type of the photoconductivity at $\lambda = 442$ nm. The sign of the detected signal corresponds to the electron one.

The frequency transfer function of the non-steady-state photo-EMF signal amplitude is shown in Fig. 3.12. The signal demonstrates typical behavior, namely, there is a linear growth of the amplitude at low frequencies of the phase modulation ω and frequency independent region at high frequencies. This regions are separated by the so-called cut-off frequency $\omega_0 = 48 - 73$ Hz. Such a behavior is an important manifestation of the adaptive nature of space charge formation in photoconductive materials, and it is described by the following expression obtained earlier for the simplest model of the semiconductor crystal with one type of partially compensated donor centers [20]:

$$J^{\omega} = \frac{Sm^2 \Delta\sigma_0 E_D}{2(1 + K^2 L_D^2)} \frac{-i\omega/\omega_0}{1 + i\omega/\omega_0}, \tag{3.24}$$

where

$$\omega_0 = [\tau_M(1 + K^2 L_D^2)]^{-1}, \tag{3.25}$$

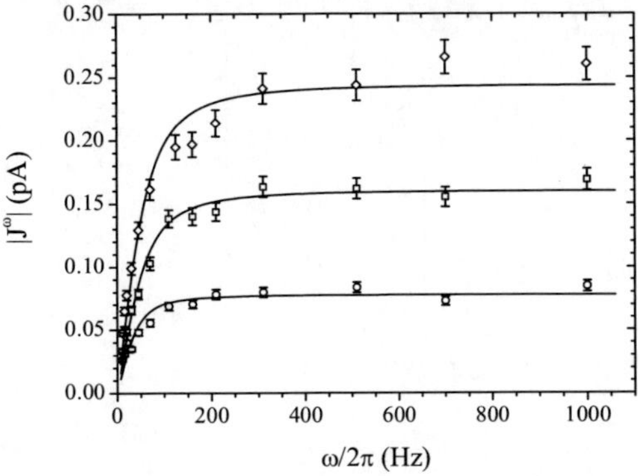

Figure 3.12. Frequency transfer functions of the non-steady-state photo-EMF measured in GaN nanostructure for different light intensities: $I_0 = 0.045$ W/cm^2 (○), 0.12 W/cm^2 (□), 0.19 W/cm^2 (◇). $\lambda = 442$ nm, $K = 7.4 \times 10^4$ cm^{-1}, $m = 0.87$, $\Delta = 0.82$. The solid lines show the approximation by Eq. (3.24).

S is the electrode area, σ_0 is the average photoconductivity of the crystal, $E_D = Kk_BT/e$ is the diffusion field, L_D is the diffusion length of photocarriers, $\tau_M = \epsilon\epsilon_0/\sigma_0$ is the Maxwell relaxation time, T is the crystal's temperature, ϵ is the dielectric constant, ϵ_0 is the free space permittivity, k_B is the Boltzmann constant, and e is the elementary charge.

If the light intensity increases we observe the corresponding growth of the signal amplitude (at $\omega > \omega_0$) and cut-off frequency (Fig. 3.13, left figure). The both dependencies demonstrate approximately linear behavior in the investigated intensity range: $J^\omega, \omega_0 \propto I_0$, what is due to increase of the sample's photoconductivity σ_0 [see Eqs. (3.24), (3.25)]. The method of the non-steady-state photo-EMF provides possibility for determination of the conductivity in high resistive materials: on measuring the cut-off frequency we can easily calculate the Maxwell relaxation time and conductivity of the sample [Eq. (3.25)]. The investigated glass has $\epsilon \simeq 4$, so the conductivity equals $\sigma = (1.1 - 1.6) \times 10^{-10}$ Ω^{-1}cm^{-1} for $I_0 = 0.045 - 0.19$ W/cm^2. One can note that cut-off frequency tends to nonzero limit at $I_0 = 0$. It is most probably associated with the finite dark conductivity of the material, which is estimated as $\sigma_d = 9 \times 10^{-11}$

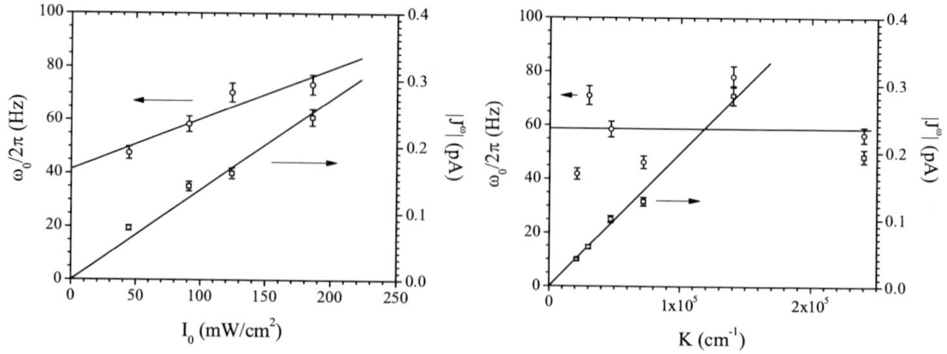

Figure 3.13. Dependencies of the non-steady-state photo-EMF amplitude (□) and cut-off frequency (○) on the intensity of incident light (left figure). $\lambda = 442$ nm, $K = 7.4 \times 10^4$ cm^{-1}. Both dependencies are approximated by linear function: $J^\omega, \omega_0 \propto I_0$. Dependencies of the non-steady-state photo-EMF amplitude (□) and cut-off frequency (○) on the spatial frequency of the interference pattern (right figure). $\lambda = 442$ nm, $I_0 = 0.19$ W/cm^2. The solid lines represent approximation by Eqs. (3.24) and (3.25) for the case of small diffusion lengths $KL_D \ll 1$.

Ω^{-1}cm^{-1}.

The dependence of the signal amplitude and cut-off frequency versus spatial frequency of the interference pattern is presented in Fig. 3.13, right figure. The dependence of the signal amplitude $|J^\omega(K)|$ seems to be linear in the range $K = 0 - 14$ μm^{-1} what is due to increase of the diffusion field E_D. The sufficient deviation of the last point from the claimed linear dependence is probably caused by the increase of the light reflection and shadowing of the near contact area but not by the blurring of the oscillating photoconductivity distribution observed at $K > L_D^{-1}$. The scattering of the cut-off frequency values is noticeable as well, and no decrease of this parameter at high spatial frequencies is guessed [see Eq. (3.25)]. So we can state that diffusion length of electrons is too small to be measured using this technique: $L_D < 40$ nm.

In this subsection we consider the excitation of the non-steady-state photo-EMF signal at $\lambda = 532$ nm and consider its temperature dependence. The sign of the signal corresponds to the electron type of the conductivity.

The frequency transfer functions of the non-steady-state photo-EMF mea-

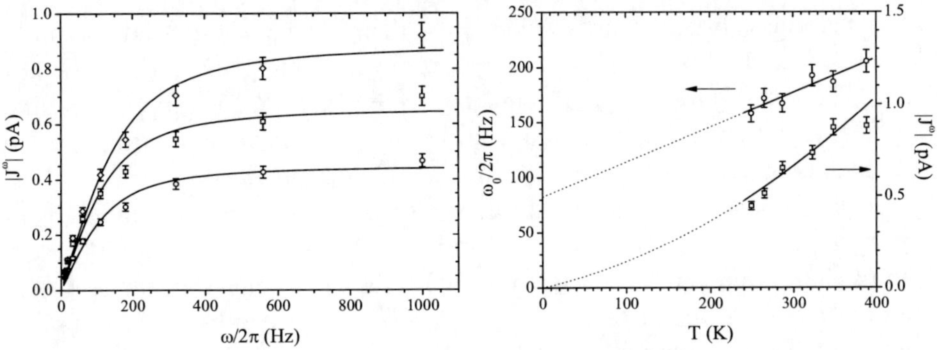

Figure 3.14. Frequency transfer functions of the non-steady-state photo-EMF measured in GaN nanostructure for different temperatures: $T = 249$ K (○), 287 K (□), 388 K (◇) (left figure). The best fit of the experimental data by Eq. (3.24) is shown by the solid lines. Dependencies of the non-steady-state photo-EMF amplitude (□) and cut-off frequency (○) on the sample temperature (right figure). The experimental dependencies $J^{\omega}(T)$ and $\omega_0(T)$ are approximated by Eqs. (3.27) and (3.26). $\lambda = 532$ nm, $I_0 = 2.3$ W/cm^2, $K = 7.1 \times 10^4$ cm^{-1}, $m = 0.91$, $\Delta = 0.61$.

sured at different temperatures are shown in Fig. 3.14, left figure. The signal demonstrates the behavior similar to the one at $\lambda = 442$ nm. The conductivity values derived from the cut-off frequencies are the following: $\sigma = (3.5 - 4.6) \times 10^{-10}\ \Omega^{-1}\text{cm}^{-1}$ ($T = 249 - 388$ K). The signal amplitude and cut-off frequency are slightly higher than the ones at $\lambda = 442$ nm but it is resulted from the much higher intensities of light so we can state that excitation of the signal at $\lambda = 532$ nm is less effective. The following reasons of this difference can be pointed out: the material have larger absorption coefficient at shorter wavelength ($\alpha = 440$ cm^{-1} at $\lambda = 442$ nm versus $\alpha = 280$ cm^{-1} at $\lambda = 532$ nm) and, perhaps, larger quantum efficiency of the conductivity.

The temperature dependence of the cut-off frequency is rather linear in the investigated range of temperatures $T = 249 - 388$ K (Fig. 3.14, right figure):

$$\omega_0 = A(T + T'), \tag{3.26}$$

where A and T' are some coefficients describing the slope and shift of the line. The theoretical dependence is unknown. We can only suppose the conductivity of the material is hopping. The temperature dependence of the conductivity in glasses is described by the conventional exponential law: $\sigma \propto \exp(-E/k_B T)$ [112]. The investigated range of temperatures is not wide enough to reveal the nonlinear growth of the cut-off frequency. So the used linear approximation can be considered as a truncated series of the unknown dependence.

Let us consider the temperature dependence of the signal amplitude measured at $\omega > \omega_0$ (Fig. 3.14, right figure). As follows from Eq. (3.24) there is an explicit temperature dependence of the signal arising from the temperature dependence of the diffusion field: $E_D \propto T$. There can be also an implicit function of the temperature resulted from the temperature dependence of material parameters, e.g. photoconductivity σ_0. The linear temperature dependence of σ_0 was assumed above, so the expected dependence of the signal amplitude becomes quadratic:

$$J^{\omega} = B(T + T')T, \tag{3.27}$$

where B is a coefficient. One can notice the saturation of the photo-EMF signal at $T > 350$ K. The nature of this saturation is not clear. Perhaps, it is due to temperature dependence of the hole conductivity: in the case of bipolar conductivity the detected signal must decrease since the stationary space charge gratings formed by electrons and holes are shifted by a half of spatial period and compensate each other [86].

Utilization of the modified technique for characterization of the as-grown structure in an external ac electric field was not successful [60]. We tried to measure a dependence of the signal amplitude on the electric field strength. No dependencies were detected in the fields up to $E_0 = 14$ kV/cm ($\Omega/2\pi \simeq 30$ kHz). This fact is explained by the small value of diffusion length which can be estimated as $L_D < k_B T/eE_0 = 18$ nm ($\mu\tau < 1.3 \times 10^{-10}$ cm^2/V). The higher amplitudes of the external field are beyond the breakdown limit which is about 14 kV/cm for our sample. If the mentioned dependence of the signal on applied field was observed, we could detect the well known dependence of the conductivity (cut-off frequency) on the frequency of applied field typical for hopping transport: $\omega_0 \propto \sigma \propto \Omega^s$. Such experiments can be considered as the task for further investigation of the effect in other nanostructured materials where transport parameters are more appropriate.

To summarize, we have investigated the effect of the non-steady-state photo-EMF in porous glass filled with nanoscaled GaN semiconductor. The detected signal revealed the behavior similar to the one typical for widegap semiconductor crystals. The conductivity of the material was estimated from the frequency transfer functions of the effect and temperature dependence of the detected signal and conductivity was observed and discussed.

4. Se within Chrysotile Asbestos

Nanowires, and particularly semiconductor nanowires, are expected to be key elements of future generation of devices used in electronics, optics, biology and medicine. Nowadays electronic circuits are close to the limit when further shrinking of elements leads to a large spread of their characteristics due to size fluctuations while the improvements of the resolution in technological equipment becomes an economical challenge [113, 114]. The utilization of nanowires as active components and interconnects seems to be perspective way to continue progress in the integration of electronic circuits. The concept of nanowire crossbar, in which active elements are placed at each crosspoint of two nanowire arrays, has been proposed [114]. It advances new types of resistive memory, reconfigurable logic circuits and new hardware realization of neuromorphic networks. The capability to assemble a wide range of different direct band gap nanowires enables the facile creation of multicolor light emitting diodes on a single substrate. Coherent laser emission has been detected from a number of nanowire cavities [115, 116]. The effect of quantum confinement inherent to nanowires should result in lower lasing thresholds, reduced temperature dependence, and a narrower gain spectrum of future devices [116]. In addition to nanoscale light sources, crossed nanowire junctions can also be configured as photodetectors critical for integrated photonics [115].

To fully explore the potential of nanowire building blocks for integrated devices and circuits, efficient and scalable strategies are needed to assemble large amount of nanowires into increasingly complex architectures. Several methods have been developed to date. They include the electric field assembly technique [117], which implies that nanowires tend to align with the direction of an applied field, fluidic flow-directed assembly technique based on passing a suspension of nanowires through a microfluidic channel formed between a mould and a flat substrate [113], and Langmuir-Blodgett assembly technique where the uniaxial compression of nanowire-surfactant monolayer on an aqueous subphase produces aligned nanowires with controlled spacing [113]. Filling dielectric matrices with desired semiconductors is one of the possible ways to get aligned nanoscaled structure [118]. The utilization of natural asbestoses allows to produce nanowires with diameter of 2-10 nm and length of 1 mm. In this paper we study composite of selenium within chrysotile asbestos for the first time.

Selenium exhibits a number of interesting and useful properties which contribute this material in investigation and production of semiconductor nanowires. Synthesis of selenium nanowires can be performed without any physical template what is due to anisotropic characteristic of the target material: the extended spiral chains of selenium atoms in the trigonal phase provide a natural template to define and guide the growth along c-axis. Several methods for selenium nanowires synthesis were proposed, they include refluxing processes [119], hydrothermal routes [120, 121], cellulose-directed growth [122], chemical vapor deposition [123], sonochemical approach [124]. Methods using physical templates have also been reported [125]: selenium nanowires can be produced by injecting chalcogene into matrix from melt under pressure or by absorption from the vapor phase.

Generally, t-Se is a p-type extrinsic semiconductor, with an indirect band gap of 1.6 eV. It has a relatively low melting point ($\sim$ 217 °C), a high photoconductivity ($\sim 0.8 \times 10^5\ \Omega^{-1}\mathrm{cm}^1$), catalytic activity toward hydration and oxidation reactions, and high piezoelectric, thermoelectric, and nonlinear optical responses. Selenium also demonstrates a high reactivity towards a wealth of chemicals that can be potentially exploited to convert selenium into other functional materials such as Bi_2Se_3, ZnSe, and CdSe.

The composition of the chrysotile asbestos is $Mg_3Si_2O_5(OH)_4$. It is a regular array of closely packed, parallel, ultrathin dielectric nanotubes. The mineral structure is described as a layer of partially hydrated MgO, which is bounded to a corresponding SiO_2. Since the lattice constant in the MgO monolayer differs from that of the SiO_2 layer, this double layer rolls up in a very thin tubes, with the outermost magnesium oxide layer. The fibers are therefore hollow cylindrical tubes with the outer diameter of 20-40 nm and the inner diameter of 2-10 nm [126, 127] that is convenient for nanowire preparation.

The investigation of the optical properties and electronic transport in nanowires and their structures is usually confined by the measurements of the light absorption [119, 121] and current-voltage characteristics [121, 128]. This approach seems to be not completely informative in the case of dissipative (non-ballistic) transport at room temperatures when scattering and even trapping of carriers occurs. The application of the standard methods for semiconductor characterization (see, for example, Ref. [44]) is often restricted by the space charge formation near the blocking contacts. This problem is of vital importance for investigation of high-resistive materials where space charge formation time can be very large and introduce inadmissible errors. The novel techniques based on holographic recording are more suitable for semi-insulating samples with low conductivity and mobility of photocarriers [2]. The expansion of the non-steady-state photo-EMF technique to the characterization of semiconductor nanowires within dielectric matrices is the main goal of this section.

We used the rather conventional experimental arrangement for the investigation of the non-steady-state photo-EMF (Fig. 1.8, b). The light from the conventional He-Ne ($\lambda = 633$ nm, $P_{out} \simeq 1$ mW) laser was split into two beams forming the interference pattern with spatial frequency K and contrast m on the sample's surface. In the most experiments the light polarization was perpendicular to the incidence plane. The signal beam was phase modulated with frequency ω and amplitude $\delta = 0.49$ by the electrooptic modulator ML-102A. Additional low frequency modulation ($\Omega/2\pi = 3$ Hz, $\Delta = 2.4$) was created by the mirror attached to the phone membrane. We studied two sample orientations: the grating vector **K** and the nanowires were parallel in the former case ("longitudinal" geometry) and they were perpendicular in the latter one ("transversal" geometry). The photocurrent arising in the sample generated the corresponding voltage on the load resistor ($R_L = 1$ MΩ for "longitudinal" geometry and $R_L = 10$ MΩ for the "transversal" one). Then this voltage was amplified and measured with the lock-in nanovoltmeter Unipan-232B or spectrum analyzers SK4-56, SK4-58.

Selenium was embedded into chrysotile asbestos matrix from the melt under pressure. Three samples were studied in this paper. Samples S1 and S2 were prepared for the measurements in the "longitudinal" geometry and sample S3 for the "transversal" one. The characteristic dimensions of the samples were $10 \times 1 \times 0.5$ mm^3. The silver paste electrodes (1×1 mm^2) were painted on the front surface (10×1 mm^2) in the "longitudinal" geometry. The electrodes (2×0.5 mm^2) were painted on the lateral surfaces in the "transversal" geometry.

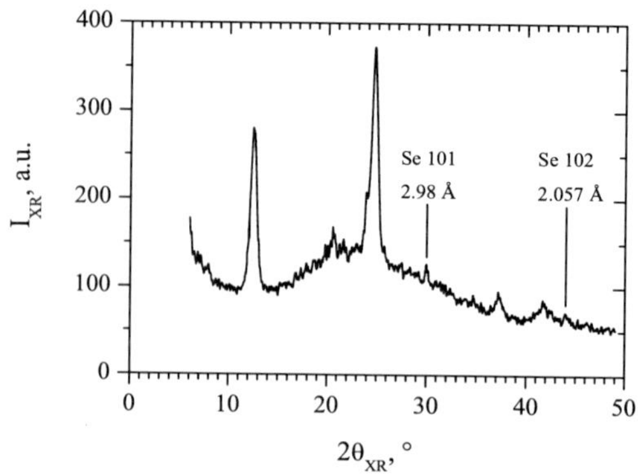

Figure 3.15. X-ray diffraction pattern of selenium within chrysotile asbestos.

The most of experiments were carried out at room temperature ($T = 16\,^{\circ}$C). In certain experiments the temperature of studied sample was changed using the Peltier thermal converter and controlled using the analog temperature-to-voltage converter.

Figure 3.15 shows XRD pattern of the studied nanocomposite. Marked weak peaks at this diagram indicate the presence of the hexagonal phase of selenium.

The presence of the non-steady-state photo-EMF effect in the new class of light-sensitive materials – quasi-1D semiconductor nanowire arrays – is the first results that should be pointed out. The amplitude of the signal detected in the experiments is of order of $0.1 - 10$ pA which is noticeable lower than that in the model objects – sillenite crystals $Bi_{12}Si(Ti,Ge)O_{20}$, where it reaches 1 nA [20]. Nevertheless this amplitude is quite enough for consistent detection of the signal with signal-to-noise ratio of $1 - 10$.

The phase of the detected signal indicates that holes prevail in the process of photo-EMF excitation. We should mention here that a nontrivial behavior of the signal amplitude was observed. In the standard experimental procedure, when there is no additional low frequency modulation, the signal amplitude is very unstable: its deviation reaches almost 100% of the mean value. These deviations are very slow: their characteristic "period" is of about 10 minutes. When additional phase modulation with frequency $\Omega/2\pi = 3$ Hz and amplitude $\Delta > 1$ is

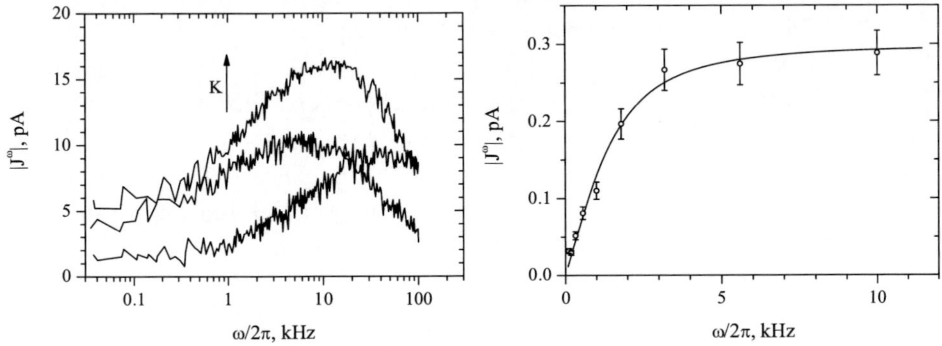

Figure 3.16. Frequency transfer functions of the non-steady-state photo-EMF signal in Se nanowires within chrysotile asbestos. The dependencies are measured for two geometries: $\mathbf{K} \parallel$ nanowires (S1), $K = 0.087\ \mu\text{m}^{-1}$, $0.26\ \mu\text{m}^{-1}$, $2.6\ \mu\text{m}^{-1}$ (left figure), and $\mathbf{K} \perp$ nanowires (S3), $K = 0.58\ \mu\text{m}^{-1}$ (right figure). $\lambda = 633$ nm, $I_0 = 370$ W/m^2, light polarization is perpendicular to the incidence plane. Solid line shows the approximation by Eq. (3.28).

switched on the signal increases by ~ 2 times, and its amplitude keeps stable within $\pm 10\%$ for hours. Such unstable behavior is most probably originated from the slow drift of the light phase in the interferometric setup. There can be two mechanisms transferring this slow drift into deviations of the detected signal. The former implies the presence of the contact photo-EMF which amplitude is comparable with that of the volume photo-EMF [130]. The latter is associated with the formation of competitive space charge gratings with different relaxation times. Since the introduction of additional modulation increases the signal amplitude regularly and for all spatial frequencies we are inclined to consider the mechanism with competitive gratings to be dominating (the contact component can both increase and decrease total signal). Anyway we settled the frequency and amplitude of additional modulation so that the influence of both factors is eliminated ($\Omega/2\pi = 3$ Hz, $\Delta = 2.4$).

We have measured the frequency transfer functions of the non-steady-state photo-EMF signal (Fig. 3.16) in the "longitudinal" and "transversal" geometries of the experiments, i.e., for the cases when direction of the space charge field (grating vector) is parallel and perpendicular to the nanowires. The signal demonstrates typical behavior, namely, there is a linear growth of the amplitude at low frequencies of the phase modulation $\omega < \omega_1$, frequency independent region at higher frequencies $\omega_1 < \omega < \omega_2$, and decay at frequencies $\omega > \omega_2$. This regions are separated by the so-called cut-off frequencies $\omega_1/2\pi = 0.42 - 12$ kHz and $\omega_2/2\pi = 32 - 210$ kHz. These dependencies are well described by the following expression which is equivalent to that obtained earlier for the simplest model of the semiconductor crystal with one type of partially compensated donor centers [38]:

$$J^\omega = J^\omega_m \frac{i\omega(\omega_2 - \omega_1)}{\omega_1\omega_2 - \omega^2 + i\omega(\omega_2 - \omega_1)}, \tag{3.28}$$

where J^ω_m is the current amplitude at the maximum of frequency dependence. The linear growth at low frequencies is an important manifestation of the adaptive nature of space charge formation in photoconductive materials. The signal in this frequency region is small since both the space charge field grating and grating of free carriers (photoconductivity grating) follow the movement of the interference pattern. The spatial shift between the gratings maintain to be nearly equal to $\pi/2$ what results in small amplitude of current. At higher frequencies (frequency independent region) the grating with larger relaxation time becomes “frozen in”, the periodic spatial shifts increase, and the photo-EMF signal reaches maximum. At very high frequencies the amplitude of oscillations for both gratings and corresponding relative shifts between them become negligible, and resulting current vanishes. Sample S2 behaves similarly to sample S1 in this and further experiments except for the temperature dependence of the effect.

The frequency transfer function measured in the “transversal” geometry looks analogously except the decay region which was not observed because of the signal smallness and corresponding frequency limits. The first cut-off frequency in this dependence equals $\omega_1 = 2.1$ kHz and it is almost the same value as that for “longitudinal” geometry (2.4 kHz).

The dependence of the signal amplitude versus spatial frequency of the interference pattern is another important characteristic that is usually measured in the non-steady-state photo-EMF experiments (Fig. 3.17). The behavior of the signal can be easily explained: the signal increases at low K due to the growth of space charge field amplitude which is proportional to the so-called diffusion

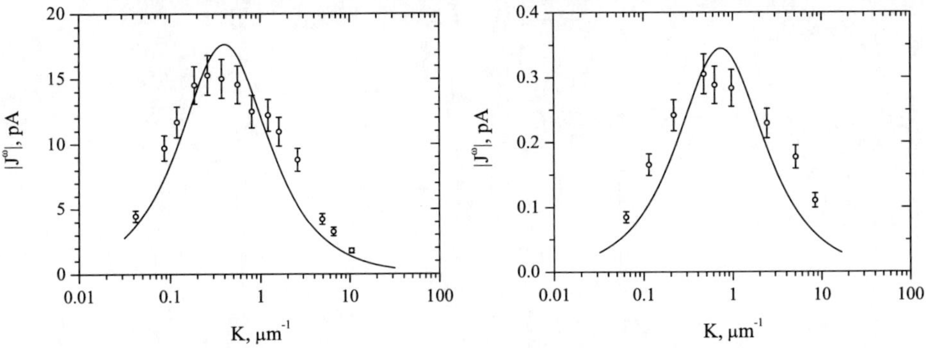

Figure 3.17. Dependence of the maximal non-steady-state photo-EMF amplitude on the spatial frequency of the interference pattern. The dependencies are measured for two geometries: **K** $\|$ nanowires (left figure), and **K** $\perp$ nanowires (right figure). $\lambda = 633$ nm, $I_0 = 370$ W/m^2, light polarization is perpendicular to the incidence plane. Solid line shows the approximation by Eq. (3.29).

field $E_d = Kk_BT/e$, the signal decrease at high K is resulted from either diffusion blurring of the conductivity grating or Debye screening of the stationary space charge grating. As the first cut-off frequency increases with spatial frequency (it will be shown below), the latter reason for the signal suppression is dominating. The dependencies are well fitted by the following simple expression known from the non-steady-state photo-EMF theory [20]:

$$J^{\omega}(K) \propto \frac{K}{1 + Q + K^2 L_S^2}. \tag{3.29}$$

Here Q is the dimensionless parameter characterizing the recombination process ($0 < Q < 1$ in general case, $Q = 0$ for linear recombination, $Q = 1$ for quadratic recombination) [40,42]. The photoconductance of the studied material demonstrated nearly square root dependence on the light intensity, i.e., quadratic recombination is present and $Q \simeq 1$. The screening length was estimated from the measured dependence: $L_S = 3.6$ μm for the "longitudinal" geometry and $L_S = 1.9$ μm for the "transversal" one.

We have also measured the dependencies of the cut-off frequencies on the spatial frequency of the interference pattern (Fig. 3.18). These dependencies noticeably differ from those measured earlier in our model objects – $Bi_{12}Si(Ti)O_{20}$

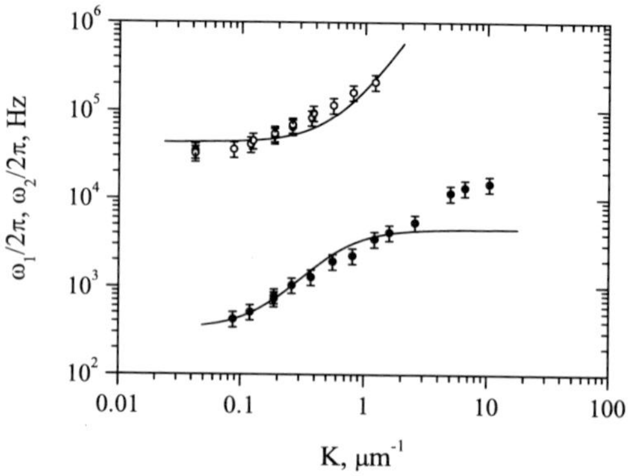

Figure 3.18. Dependencies of the first (•) and second (∘) cut-off frequencies on the spatial frequency of the interference pattern. $\lambda = 633$ nm, $I_0 = 370$ W/m^2, $\mathbf{K} \parallel$ nanowires, light polarization is perpendicular to the incidence plane. Solid lines show the approximation by Eqs. (3.30) and (3.31).

crystals. Dependence $\omega_1(K)$ is expected to be decreasing function in particular. The expressions for the cut-off frequencies obtained earlier for the simplest model of semiconductor with one type of compensated donor centers are as follows [38]:

$$\omega_1 = \frac{1 + Q + K^2 L_S^2}{\tau + \tau_M(1 + Q + K^2 L_d^2)}, \tag{3.30}$$

$$\omega_2 = \omega_1 + \frac{1}{\tau_M} + \frac{1 + Q + K^2 L_d^2}{\tau}. \tag{3.31}$$

The only explanation in the frames of existing theory can be done for the observed growing function $\omega_1(K)$, if we do not neglect the screening effects: $L_S^2 > L_d^2 \tau_M/(\tau + \tau_M)$. The following parameters were determined from the experimental dependencies: $\tau = 0.96$ ms, $\tau_M = 3.7$ μs, $L_d = 27$ μm, $L_S = 8.9$ μm. As seen from Fig. 3.18 the correspondence of the experimental and theoretical dependencies $\omega_1(K)$ is just qualitative: they noticeably diverge at high spatial frequencies.

The anisotropy of the studied material reveals not only in the difference of the current amplitude along and across nanowires (Fig. 3.16) but also in the

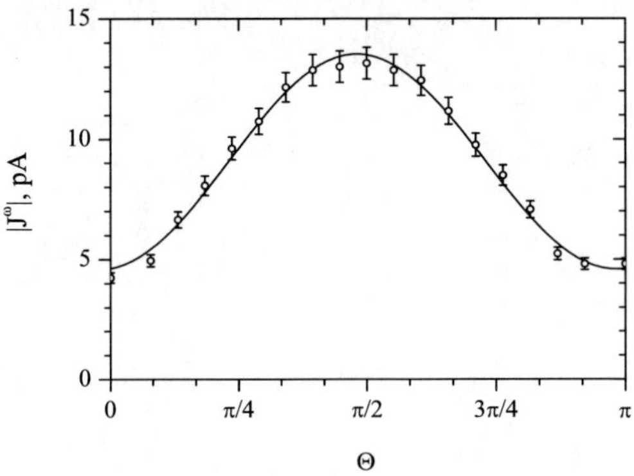

Figure 3.19. Dependence of the non-steady-state photo-EMF amplitude on the angle of light polarization. Angle $\Theta = 0$ corresponds to the case when light is polarized in the incidence plane (polarization is nearly parallel to the nanowires). $\lambda = 633$ nm, $I_0 = 370$ W/m^2, $K = 0.20\ \mu\text{m}^{-1}$, $\omega/2\pi = 35$ kHz, $\mathbf{K} \parallel$ nanowires.

experiments with varied light polarization (Fig. 3.19). The higher signal amplitudes are observed in the case when the light is polarized perpendicular to nanowires. The influence of mobility anisotropy is eliminated in this part of experiments since the current always flow along the nanowires, so the difference in signal amplitude should be attributed to the photoexcitation processes. Such photoexcitation anisotropy is probably associated with the orientation of selenium crystallites with respect to the matrix: dipole momentum of Se is perpendicular to asbestos nanotubes.

We have mentioned above that the the signal nontrivially behave in the presence of additional low-frequency modulation. The dependencies of the non-steady-state photo-EMF signal versus amplitude of additional modulation is presented in Fig. 3.20. The monotonous growth of the signal amplitude in this dependence is rather unexpected. The basic non-steady-state photo-EMF theory predicts the signal decay as $(1 + \Omega\Delta/\omega_1)^{-0.5}$ [129]. It was pronounced in the beginning of the section that the grating competition is responsible for such behavior. At the first glance two mechanisms for grating competition can be suggested: influence of shallow traps and bipolar conductivity. The analysis

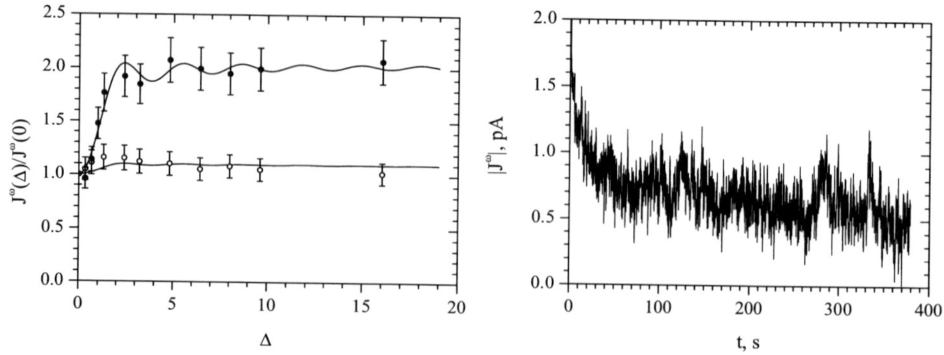

Figure 3.20. Dependencies of the non-steady-state photo-EMF amplitude on the amplitude of additional low-frequency phase modulation ($\Omega/2\pi = 3$ Hz) (left figure). (○) – $K = 0.099\ \mu\text{m}^{-1}$, $\omega/2\pi = 9.0$ kHz; (•) – $K = 12\ \mu\text{m}^{-1}$, $\omega/2\pi = 59$ kHz. Time evolution of the signal amplitude after the moment when additional phase modulation ($\Omega/2\pi = 3$ Hz, $\Delta = 16$) is switched off (right figure). $K = 12\ \mu\text{m}^{-1}$, $\omega/2\pi = 59$ kHz, integration time $\tau_{int} = 0.3$ s. $\lambda = 633$ nm, $I_0 = 370$ W/m^2, $\mathbf{K} \parallel$ nanowires, light polarization is perpendicular to the incidence plane.

of the effect for the model with shallow traps is presented in the next section. In this case the charge gratings with the opposite signs are formed at deep donor and shallow levels. The solid lines in Fig. 3.20 represent approximation by Eq. (3.52) with fitting parameter $\beta T = 0.91$ for $K = 0.099\ \mu\text{m}^{-1}$, $\omega/2\pi = 9.0$ kHz and $\beta T = 0.49$ for $K = 12\ \mu\text{m}^{-1}$, $\omega/2\pi = 59$ kHz. The consideration of the effect in the presence of bipolar conductivity was not successful: the desired theoretical dependence is not obtained (at least for the simplest model).

The relaxation of the signal amplitude between the regimes with and without additional phase modulation occurs during considerable time amount (Fig. 3.20). We have approximated this dependence by the simplest exponential decay function and found the time constant $\tau_{am} = 35$ s as a fitting parameter.

The temperature dependence of the non-steady-state photo-EMF was also observed (Fig. 3.21). The two samples prepared for measurements in the "longitudinal" geometry demonstrate different behavior in the range of room temperatures. The non-steady-state photo-EMF signal in sample S1 linearly grows

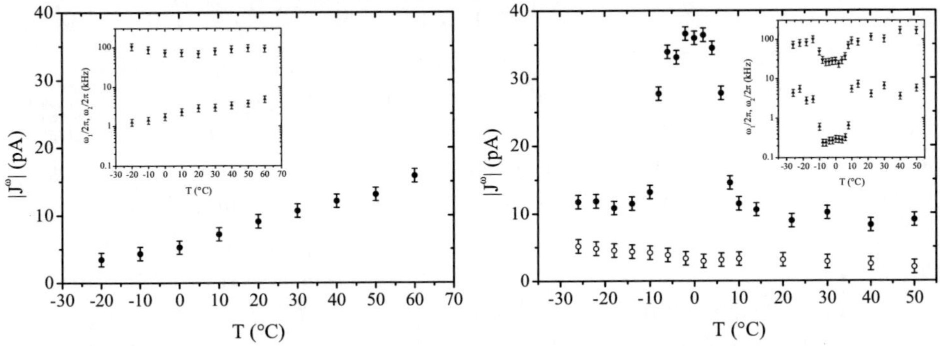

Figure 3.21. Temperature dependencies of the non-steady-state photo-EMF amplitude measured in sample S1 (left figure) at $K = 0.36\ \mu\text{m}^{-1}$ and in sample S2 (right figure) at $K = 0.11\ \mu\text{m}^{-1}$ (•) and $K = 7.2\ \mu\text{m}^{-1}$ (∘). The insets present temperature dependencies of the first (▲) and the second (▼) cut-off frequencies. $\lambda = 633$ nm, $I_0 = 370$ W/m^2, **K** $\parallel$ nanowires, light polarization is perpendicular to the incidence plane.

with the temperature. The linear growth also follows from the theoretical approach which implies that the signal amplitude is proportional to the diffusion field $E_d = Kk_BT/e$. The slope of the experimental dependence is larger however: being extrapolated to the lower temperatures it crosses the axis of abscissae at $T = -38$ °C (235 K) but not at $T = 0$ K as follows from the simplest theory. This difference can be explained by the temperature of the material parameters (e.g., conductivity). The temperature dependence of the cut-off frequencies proves this assumption. So the observed linear dependence in studied region should be considered as a linear approximation of more complicated temperature dependence. An unexpected temperature dependence of the effect was revealed in sample S2. There are an pronounced peak in the dependence of the signal amplitude and wells in the dependencies of the both cut-off frequencies measured at low K. These peculiarities appear at $T \simeq 0$ °C. In this case we can hardly attribute such behavior to the change of material parameters: such assumption requires simultaneous growth of the photoconductivity σ_0, Maxwell relaxation time τ_M, and conductivity relaxation time τ, but $\tau_M \propto \sigma_0^{-1}$. The more realistic explanation implies that dc electric field appears when temperature approaches to 0 °C. It is known from the basic theory of the non-steady-state photo-EMF effect that the amplitude of photocurrent grows in the presence of external dc field and its maximum at the frequency transfer function shifts to the lower frequencies [38]. We performed measurements of the dc electric current arising in the samples under uniform illumination and without it. The temperature dependencies of this dc current demonstrate peak in the vicinity of 0 ° in both samples S1 and S2. The peak in sample S1 is of 10 pA only, and in sample S2 it reaches 200 pA. We should note here that this dc current is almost independent on the illumination level, so it is rather pyroelectric than photogalvanic. The origin of this pyroelectricity is not clear, but evidently it is associated with the crystallization of atmosphere water accumulated in the sample pores unfilled with selenium.

The studied material belongs to the class of quasi one-dimensional noncrystalline media. The most of obtained experimental results, however, are well described by the theory developed for bulk crystals. The dependence of signal amplitude versus amplitude of additional phase modulation is the only result that seriously contradicts with the theory developed earlier for the simplest one-level model of semiconductor crystal [129, 130]. It was mentioned that the unexpected behavior can be associated with the competition of space charge gratings characterized by different relaxation times. Grating competition is one of the feature of the two-level model of semiconductor [41]. There is no theory of the non-steady-state photo-EMF in the presence of additional modulation for this model. So we shall try to explain this certain feature of the signal behavior using the approach based on band transport two-level model of semiconductor [41, 44]. For the simplicity the material is assumed to be isotropic as well.

The initial stage of the analysis is similar to that described above in Section 1.1. Let us suppose the sample is illuminated by the interference pattern formed by two plain light waves one of which is phase modulated:

$$I(x,t) = I_0\{1 + m\cos[Kx + \varphi(t)]\}, \tag{3.32}$$
$$\varphi(t) = \delta\cos(\omega t) + \Delta\cos(\Omega t). \tag{3.33}$$

Here I_0, m, and K are the average light intensity, contrast, and spatial frequency of the interference pattern, Δ and ω are the amplitude and frequency of phase modulation. The non-steady-state photo-EMF signal is defined by the drift component of the electric current averaged over interelectrode spacing [Eq. (1.17)].

Necessary distributions $n(x,t)$ and $E(x,t)$ are derived from Eqs. (1.1)-(1.5).

The calculation procedure simplifies substantially if we suppose the contrast m to be small: $m \ll 1$. Then we can look for the solution of equation set to the lowest order with respect to small parameter m presenting $n(x,t)$, $E(x,t)$ as the sums of sinusoidal gratings:

$$n(x,t) = n_0 + n_+(t)\exp(iKx) + n_-(t)\exp(-iKx), \qquad (3.34)$$

$$E(x,t) = E_+(t)\exp(iKx) + E_-(t)\exp(-iKx). \qquad (3.35)$$

Furthermore, we suppose the amplitude of primary phase modulation to be small $\delta \ll 1$, and thus we can present complex amplitudes n_+, n_-, E_+, E_- like the following series:

$$n_+(t) = \sum_{p=-1}^{1} \sum_{q=-\infty}^{\infty} n_{+pq} \exp[i(p\omega + q\Omega)t]. \qquad (3.36)$$

Then the complex amplitude of the nonstationary photocurrent with frequency ω can be rewritten as the following combination of coefficients:

$$j^{\omega} = 2e\mu \sum_{q=-\infty}^{\infty} \left(n_{+1q}E_{-0-q} + n_{-0-q}E_{+1q} + n_{-1q}E_{+0-q} + n_{+0-q}E_{-1q}\right). \qquad (3.37)$$

After substitution of Eqs. (3.32)-(3.37) into Eqs. (1.1)-(1.5), linearization and some routine calculations we obtain rather common expression for the amplitude of the non-steady-state photo-EMF signal:

$$j^{\omega} = \frac{m^2\delta}{2}\sigma_0 E_d \sum_{q=-\infty}^{\infty} \mathrm{J}_q^2(\Delta)\frac{A(\omega, q\Omega)}{B(-q\Omega)B(\omega+q\Omega)}, \qquad (3.38)$$

where

$$\begin{aligned} A(\omega, q\Omega) &= -\tau_M\tau_T^2[iq\Omega(\omega^2 + 3\omega q\Omega + 2q^2\Omega^2) \\ &\quad + i(\omega + 2q\Omega)(\beta + \tau_{TI}^{-1})(i\omega + \beta + \tau_{TI}^{-1})], \qquad (3.39) \\ B(\omega) &= -i\omega^3\tau_M\tau_D\tau_T - \omega^2[\tau_M\tau_D\tau_T(S_D I_0 + \tau_{DI}^{-1} + \beta + \tau_{TI}^{-1}) \\ &\quad + \tau_M(\tau_D + \tau_T) + \tau_D\tau_T(1 + K^2L_M^2)] + i\omega[\tau_D + \tau_T \end{aligned}$$

$$
\begin{aligned}
&+\tau_D(S_D I_0+\tau_{DI}^{-1})(\tau_T+\tau_M(1+K^2L_T^2))+\tau_T(\beta+\tau_{TI}^{-1})\\
&\times(\tau_D+\tau_M(1+K^2L_D^2))+\tau_M\tau_D\tau_T(S_D I_0+\tau_{DI}^{-1})(\beta+\tau_{TI}^{-1})]\\
&+\tau_D(S_D I_0+\tau_{DI}^{-1})+\tau_T(\beta+\tau_{TI}^{-1})\\
&+\tau_D\tau_T(S_D I_0+\tau_{DI}^{-1})(\beta+\tau_{TI}^{-1})(1+K^2L_M^2),
\end{aligned}
\tag{3.40}
$$

$E_d = Kk_BT_s/e$ is the diffusion field, k_B is the Boltzmann constant, T_s is the sample's temperature, $L_D = \sqrt{D\tau_D}$, $L_T = \sqrt{D\tau_T}$ and $L_M = \sqrt{D\tau_M}$ correspond to the average diffusion shifts of electrons for times τ_D, τ_T and τ_M, respectively, $\mathrm{J}_q(\Delta)$ is the Bessel function of the first kind of the q-th order.

Equations (3.38)-(3.40) are evidently too complicated for the further analysis, so let us simplify them using the following assumptions. We suppose the light intensity to be not very high so that the deep donors keep almost filled:

$$N_D \gg N_A + N_0. \tag{3.41}$$

The shallow traps are assumed to be almost empty on the contrary:

$$M_0 \ll M_T. \tag{3.42}$$

The low illumination level does not allow to produce high density of electrons in the conduction band:

$$n_0 \ll N_0, M_0. \tag{3.43}$$

Other set of conditions concern the characteristic timescales and frequencies of phase modulation. First let us suppose that the frequency of primary modulation is of order of τ_M^{-1}, while the frequency of additional modulation is much lower: it is chosen with respect to the rate of slow relaxation processes at shallow traps. So the following conditions are valid:

$$\omega \gg \Omega\Delta, \tag{3.44}$$

$$\omega \gg \beta, \tag{3.45}$$

$$\tau_M \ll \beta^{-1}. \tag{3.46}$$

Another condition establish the relation between the Maxwell relaxation time and lifetime of ionized donors and it is often present in semiinsulating materials:

$$\tau_M \ll \tau_{DI}. \tag{3.47}$$

At last, the spatial frequency of interference pattern is assumed to be not very high:

$$K^2 L_M^2 \ll M_0/n_0. \tag{3.48}$$

If conditions (3.41)-(3.48) are fulfilled, the expression for the complex amplitude of the non-steady-state photo-EMF signal reduces to the form very similar to that known from basic theory of the effect:

$$j^\omega = \frac{-i\omega\tau_M(m^2\delta/2)\sigma_0 E_d}{1-\omega^2\tau\tau_M + i\omega[\tau+\tau_M(1+K^2L_d^2)]}\frac{N(\Delta,\Omega)}{(1+\tau_D/\tau_T)(1+\tau_D/\beta\tau_T\tau_{DI})} \tag{3.49}$$

where $\tau = \left(\tau_D^{-1}+\tau_T^{-1}\right)^{-1}$, $L_d = \sqrt{D\tau}$ are the lifetime and diffusion length of electrons, respectively, and

$$N(\Delta,\Omega) = \mathrm{J}_0^2(\Delta) + 2\sum_{q=1}^{\Delta_m} \mathrm{J}_q^2(\Delta)\frac{1+(q\Omega)^2\beta^{-1}T}{1+(q\Omega T)^2} \tag{3.50}$$

is the normalized non-steady-state photo-EMF amplitude: $N = j^\omega(\Delta)/j^\omega(0)$ – the main goal of this analysis, with characteristic time

$$T = \frac{\tau_D+\tau_T}{\tau_D\tau_{DI}^{-1}+\beta\tau_T}. \tag{3.51}$$

The upper limit of sum Δ_m is chosen to be equal to the nearest integer greater than amplitude Δ.

Function $N(\Delta,\Omega)$ is real and it can be both growing and decaying depending on relation between material parameters β, T and frequency Ω. It should be noted that characteristic time T is rather large: the denominator in Eq. (3.51) can be rewritten as $n_0/(N_A+N_0)+n_0/M_0$, and according to Eq. (3.43) it makes time T much larger than sum $\tau_D+\tau_T$. So frequency of additional modulation Ω remaining low in sense of Eq. (3.44) can be chosen to satisfy conditions $(\Omega T)^2 \gg 1$ and $\Omega^2\beta^{-1}T \gg 1$. In this case the expression for normalized amplitude N can be further simplified:

$$N(\Delta) = \mathrm{J}_0^2(\Delta) + \frac{2}{\beta T}\sum_{q=1}^{\Delta_m}\mathrm{J}_q^2(\Delta). \tag{3.52}$$

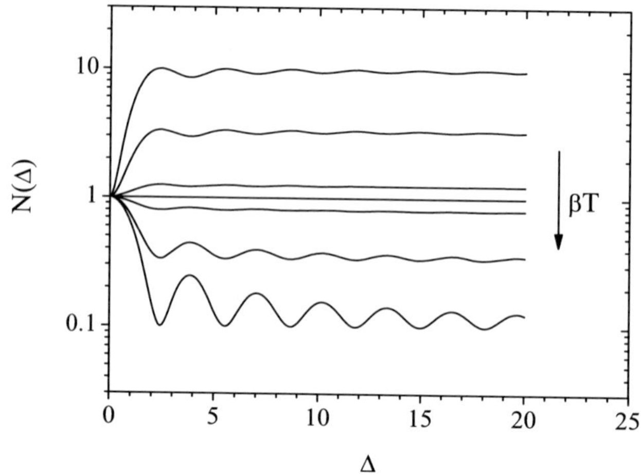

Figure 3.22. Normalized amplitude of the non-steady-state photo-EMF N versus amplitude of additional low-frequency phase modulation Δ. The dependencies are calculated using Eq. (3.52) for $\beta T = 0.1, 0.3, 0.8, 1.0, 1.25, 3.0$, 10.

As seen, this function has the only argument – the amplitude of additional phase modulation Δ, and it contains the only parameter – product βT. Dependence $N(\Delta)$ calculated for different values of parameter βT is presented in Fig. 3.22. The dependence acquires the growing character at $\beta T < 1$ and decaying one at $\beta T > 1$. The growth (decay) at low amplitudes is nearly quadratic: $N \propto \Delta^2$. The saturation level is reached at $\Delta \simeq 2$ and equals $(\beta T)^{-1}$.

To summarize we studied the effect of non-steady-state photo-EMF in new class of composite materials – nanostructured selenium within chrysotile asbestos. The average photoconductivity of the material was estimated from the measurements of the signal amplitude versus frequency of phase modulation. The dependence of the signal on spatial frequency provided the estimation of the diffusion and Debye screening lengths. The polarization dependencies revealed the anisotropy of photoexcitation while the temperature dependence of the detected signal pointed out to the presence of water in the porous matrix. The increase of the signal amplitude was observed on application of additional low-frequency phase modulation. To explain this peculiarity the theoretical analysis has been performed for the model of semiconductor with shallow traps.

The obtained results have demonstrated that the non-steady-state photo-EMF technique can be considered as a powerful tool in investigation of nanostructured media. A number of photoelectric parameters are estimated using the only arrangement which can be easily automatized in the case of need. The technological problem of ohmic contacts to widegap semiconductors, especially nanosized ones, is not so significant here as that for standard methods, since the effect of the non-steady-state photo-EMF is based on the excitation of an alternating current. The developed theoretical approach is not restricted to considered material and can be applied to any semiconductor material (bulk or nanostructured) where electronic transport is accompanied by the charge trapping. The behavior of the non-steady-state photo-EMF in semiconductor nanowires at low temperatures, where ballistic transport and Coulomb blockade effects can be present, as well as the application of such materials in adaptive detectors of phase-modulated optical signals seem to be interesting future tasks.

5. Neutron Irradiated SiC

Space charge formation in photoconductive and photorefractive media is accompanied by an electric current, which redistributes carriers between bright and dark regions [2]. This idea has been implemented in the so-called effect of nonstationary holographic currents (or non-steady-state photo-EMF) arising via dynamic space charge gratings [19, 20, 38]. The effect reveals itself as an alternating electric current arising in a semiconductor illuminated by an oscillating interference pattern. The related techniques realized with an interference pattern running with constant velocity are called as holographic current [131] or moving photocarrier grating technique [32]. The electric current appears due to spatial shifts of the space-charge and photoconductivity distributions in the crystal volume.

The excitation of the non-steady-state photo-EMF signal involves processes of carrier generation, diffusion and drift in an electric field, recombination to local centers. This complex nature of the effect makes it a powerful tool for determination of photoelectric parameters such as type and value of photoconductivity, carrier lifetime, mobility and diffusion length, density of recombination centers, etc. [32, 38, 132–134]. Besides tasks of material characterization the nonstationary holographic currents are used for the detection of phase- and frequency-modulated optical signals [21, 130, 135–142].

In this section we shall describe application of the non-steady-state photo-EMF technique for investigations of a SiC crystal. Silicon carbide is regarded as perspective material for opto- and power electronics [143–147]. High thermal, chemical and radiation stability allows its utilization in the production of ionizing radiation detectors [148–150]. The space charge waves have already been studied in undoped SiC crystals [151, 152]. In contrast to the mentioned approach we shall investigate the nonresonant techniques of photo-EMF excitation in the silicone carbide crystal preliminarily irradiated by reactor neutrons.

The experiments on excitation of nonstationary holographic currents in SiC are carried out with the arrangements used earlier for investigations of other wide-bandgap semiconductors and nanostructured materials [60, 85, 100, 107, 153–155]. The second harmonic of Nd:YAG laser ($\lambda = 532$ nm, $P_{out} \simeq 20$ mW) is split into two beams, which then create the interference pattern with spatial frequency K, contrast m and average intensity I_0 on the crystal surface ($m = 0.45$ and 0.98 in the experiments with and without external field, respectively). The electro-optic modulator introduces phase modulation with amplitude $\delta = 0.61$ and frequency ω into the signal beam. The polarization plane is perpendicular to the incidence plane. The photocurrent arising in the sample produces a voltage across the load resistor, which was amplified and then measured by spectrum analyzers SK4-56, SK4-58. The sinusoidal voltage with frequency $\Omega/2\pi = 100$ kHz is applied to the sample in experiments with external field.

Additional investigations of the photoconductivity relaxation are performed using the standard technique described, for example, in Ref. [156]. This technique includes measurements of frequency characteristic of photoconductivity response to the amplitude-modulated light. The sinusoidal modulation of light intensity is created by the electro-optic modulator (with modulation depth $m_{AM} = 0.35$). A dc voltage of 400 V is applied to the sample (this voltage is enough for the breakdown of the blocking electrode).

The crystal $6H$-SiC was grown by the sublimation sandwich method [157]. The growth was carried out in an argon atmosphere at 2400 °C. The obtained crystals contains nitrogen donor impurity as well as acceptor impurities of aluminium and boron with concentration of $\sim 10^{17}$ cm^{-3}. The samples possess the

Table 3.1. Parameters used for the calculation of dependencies in Fig. 3.23

	I_0, mW/cm^2	
	7.1	570
G_0, Ω^{-1}	3.1×10^{-10}	1.1×10^{-8}
τ_D, s	3.6×10^{-4}	1.2×10^{-4}
τ_T, s	1.4×10^{-4}	6.9×10^{-5}
B, s^{-1}	8.1×10^2	1.4×10^3
σ_0, Ω^{-1}cm^{-1}	2×10^{-8}	7×10^{-7}

n-type of conductivity with electron density of $(5-10) \times 10^{17}$ cm^{-3}. The crystals were ground and polished to the thickness of ~ 0.5 mm, and then they were etched in KOH alkali melt at 480 °C in order to remove mechanically damaged surface layer and estimate the density of protruding dislocations (10^3 cm^{-2}).

Afterwards the crystals were irradiated by reactor neutrons at temperature of 100 – 150 °C. The energy distribution of reactor neutrons is very wide, the fluxes of thermal and fast neutrons were of 5.5×10^{13} cm^{-2}s^{-1} and 3.8×10^{12} cm^{-2}s^{-1}, respectively. The radiation fluence was of 10^{17} cm^{-2}.

The sample has dimensions of $6.0 \times 1.0 \times 0.47$ mm. The front and back surfaces (6.0×1.0 mm), which are (0001) crystal cut, have the optical quality. The silver paste electrodes are deposited on the lateral surfaces (6.0×0.47 mm).

It is reasonable to start consideration of the crystal's properties using the standard techniques based on the conductivity measurements. First, we measured the conductivity of initial non-irradiated crystal. We can state, that such crystals have a rather high dark conductivity $\sigma_d \simeq 3 \times 10^{-3}$ Ω^{-1}cm^{-1}, while the photoconductivity do not exceed $\sigma_0 < 10^{-6}$ Ω^{-1}cm^{-1} for $\lambda = 532$ nm and $I_0 = 200$ mW/cm^2 (the estimations are obtained by conventional measurements of the sample resistance). The relation $\sigma_0/\sigma_d < 10^{-3}$ points out to the uselessness of initial crystals for experiments with dynamic space charge gratings (the grating amplitude is small in such media and contact signals, on the contrary, are large). Inverse relation $\sigma_0 \gg \sigma_d$ is observed in the neutron irradiated crystal, and namely this crystal is of interest for us.

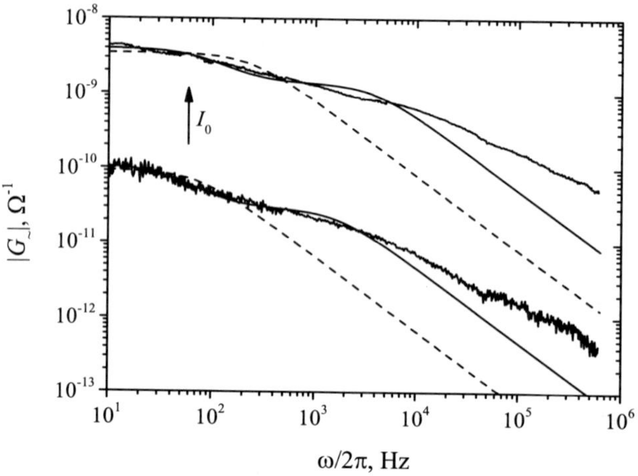

Figure 3.23. Frequency dependencies of the photoconductance response to amplitude modulated light. $I_0 = 7.1, 570$ mW/cm^2. The green dash and blue solid lines show approximations for one-level and two-level models of a semiconductor, respectively

Frequency transfer functions of the photoconductivity response are shown in Fig. 3.23. We have tried to approximate the dependencies using the approach developed for the simplest model of semiconductor with one type of partially compensated donor centers [156]:

$$G_{\sim}(\omega) = \frac{m_{AM} G_0}{1 + i\omega\tau}, \tag{3.53}$$

where G_0 is the sample photoconductance and τ is the photoconductivity relaxation time, which is equal to the lifetime of photocarriers in the simplest case. The fitting parameters G_0 and τ were estimated: $G_0 = (0.27, 9.8) \times 10^{-9}\ \Omega^{-1}$ and $\tau = 2.2$, 0.67 ms for the light intensities $I_0 = 7.1$, 570 mW/cm^2. As seen the experimental and calculated curves significantly differs at large frequencies (Fig. 3.23, dash lines). This difference can be noticeably reduced using the two-level model of semiconductor, where electron excitation occurs from both deep donors and shallow traps [156]. In the case of linear generation and recombination for deep levels and low population of shallow levels we can still use Eq.

(3.53) with a frequency-dependent time τ [40]:

$$\tau(\omega) = \tau_D \left(1 + \frac{\tau_T^{-1} B^{-1}}{1 + i\omega B^{-1}}\right), \tag{3.54}$$

where τ_D, τ_T are the electron lifetimes with respect to the deep donors and shallow traps, respectively, $B = \beta + S_T I_0$ is the total excitation rate from shallow traps, β is the thermal excitation rate from shallow traps, S_T is the ionization cross section for shallow traps. The approximation using Eqs. (3.53) and (3.54) allows us to estimate a number of crystal parameters (Table 3.1). The photoconductivity σ_0 is calculated from values G_0 taking into account the geometrical sizes of the sample, diameter of the light beam and the absorption coefficient $\alpha = 60$ cm^{-1}. A comparison of σ_0 values for two light intensities points out to the small nonlinearity of the photoconductivity: $\sigma_0 \propto I_0^{0.82}$. The agreement between experimental and calculated curves is still poor at high frequencies. The utilization of the more complex multilevel model of semiconductor can solve this problem, but this step will produce numerous fitting parameters.

Now let us proceed with the experiments on nonstationary holographic currents. The signal amplitude is of $10^{-12} - 10^{-8}$ A, what is nearly equal to the values observed in our model objects – sillenite crystals $Bi_{12}Si(Ti,Ge)O_{20}$ [20, 60, 153, 155]. Such amplitudes are enough for the detection with signal-to-noise ratio of $0 - 80$ dB. The phase of the signal corresponds to the electron type of photoconductivity.

We have measured frequency transfer functions of the holographic current in as fabricated SiC crystal (Fig. 3.24). The dependencies contain the growth of the signal at low ω followed by the frequency independent region, as usual. The signal growth at low frequencies is an important manifestation of the adaptivity of space charge formation in photoconductive materials [2, 20]. The signal is small because of the fact, that both the space charge field grating and the grating of free electrons (photoconductivity grating) follow movements of the interference pattern. The spatial shift between them is nearly equal to $\pi/2$, what finally results in small value of the average drift component of the current. The grating with larger relaxation time becomes "frozen-in" at higher frequencies, the spatial shifts between gratings increase, and holographic current reaches its maximum (frequency independent region). When the region of decaying signal is present, it is usually attributed to the finiteness of carriers lifetime τ, and the

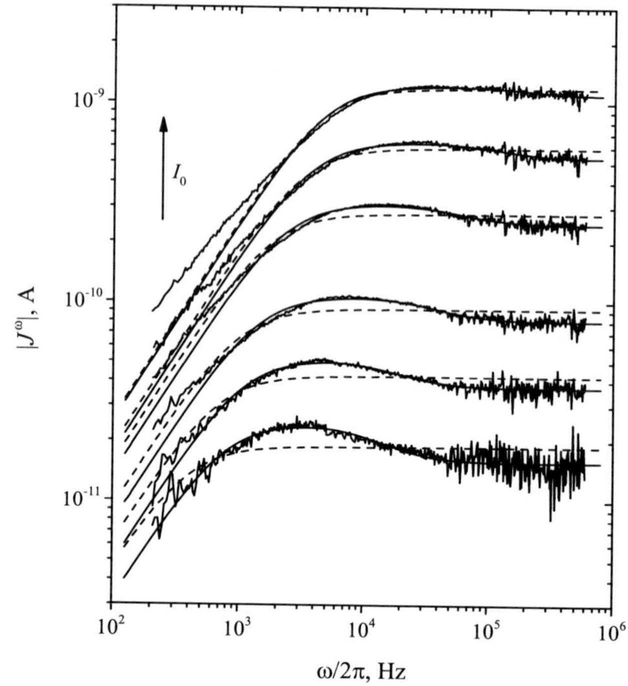

Figure 3.24. Frequency transfer functions of the non-steady-state photo-EMF measured for different light intensities $I_0 = 59, 110, 210, 540, 1100, 2200$ mW/cm^2. $K = 0.36\ \mu\text{m}^{-1}$. The green dash and blue solid lines show approximations for one-level and two-level models of semiconductor [Eq. (3.55) with $\tau = \text{const}$ and $\tau = \tau(\omega)$, respectively].

signal behavior is described by the following expression [38]:

$$j^{\omega} = \frac{-(m^2\Delta/2)\sigma_0 E_D i\omega\tau_M}{1 - \omega^2\tau\tau_M + i\omega[\tau + \tau_M(1 + K^2L_D^2)]}. \tag{3.55}$$

Here $E_D = (k_BT/e)K$ is the diffusion field, $\tau_M = \epsilon_0\epsilon/\sigma_0$ is the Maxwell relaxation time, L_D is the diffusion length of electrons [2]. Approximation of the experimental curves by this expression turns out to be not very successful (Fig. 3.24, dashed lines).

The description using the two-level model of semiconductor seems to be more appropriate (Fig. 3.24, solid lines). We have applied the same Eq. (3.55) but with the frequency-dependent effective lifetime $\tau(\omega)$ [40], as it was done for the photoconductivity response. Since there is the only region where the current decays, the expression for $\tau(\omega)$ can be simplified:

$$\tau(\omega) = \frac{\tau_D \tau_T^{-1} B^{-1}}{1 + i\omega B^{-1}}. \qquad (3.56)$$

Values τ_M, τ_D/τ_T and B are fitting parameters for the frequency transfer functions measured at low K (Fig. 3.24). These parameters vary with the light intensity, and these dependencies are considered below. In fact, there is the fourth fitting parameter $S_{eff}(m^2\Delta/2)\sigma_0 E_D$, which is just the proportionality coefficient defining the signal amplitude. We usually skip its notation, since it includes effective cross-section area of the sample S_{eff}, which depends on the area and geometry of electrodes, distribution of the average light intensity in the beams (gaussian) and light absorption.

Frequency dependence with two frequency independent regions was observed earlier in crystal CdTe:Ge [134]. Such behavior was also associated with the presence of two types of photoactive centers, which make relaxation of charge distribution to be double-stage.

The dependencies of the signal amplitude and first cut-off frequency versus illumination level are presented in Fig. 3.25. The amplitude of the photocurrent in the maximum of frequency transfer function is proportional to the light intensity: $J^\omega \propto I_0$. There is, however, a sublinear dependence of the cut-off frequency (which is defined by the decay of $\sqrt{2}$ from the maximum signal amplitude): $\omega_1 \propto I_0^{0.50}$. This nonlinearity can be due to both the corresponding nonlinearity of the photoconductivity and the fact, that the cut-off frequency if defined by the combination of times τ and τ_M in crystals with large electron lifetimes [38, 40]. Moreover, the presence of dark conductivity can contribute in the initial stage of dependence $\omega_1(I_0)$.

Generally speaking, the phenomenological description using a cut-off frequency ω_1 is not informative enough for the studied SiC crystal with its nontrivial behavior of the photoconductivity. As it was mentioned above, the frequency dependencies are well fitted in the frames of two-level model of semiconductor with the parameter set τ_M, τ_D/τ_T, B. Dependencies of these parameters

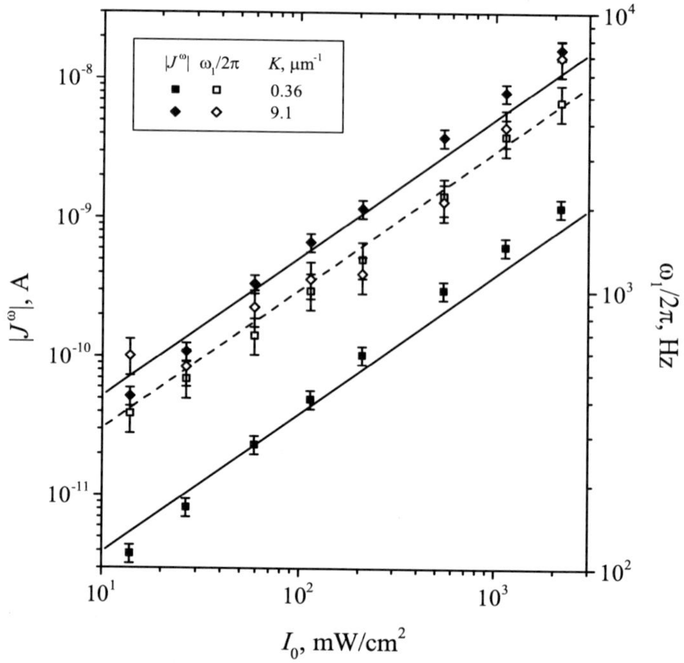

Figure 3.25. Dependencies of the signal amplitude and first cut-off frequency on the average light intensity. $K = 0.36, 9.1\ \mu\text{m}^{-1}$.

on the light intensity are shown in Fig. 3.26. Dependence $\tau_M \propto I_0^{-0.51}$ indicates the nonlinearity of the electron photogeneration and recombination in our sample. Values of τ_M correspond to the following range of conductivity: $\sigma_0 = (2.1 - 26) \times 10^{-9}\ \Omega^{-1}\text{cm}^{-1}$. There is a weak intensity dependence of τ_D/τ_T parameter: $\tau_D/\tau_T \propto I_0^{-0.41}$. Parameter B demonstrate the expected dependence on light intensity: $B(I_0) = \beta + S_T I_0$ with thermal generation rate $\beta = 8.0 \times 10^4\ \text{s}^{-1}$ and ionization cross section $S_T = 6.6 \times 10^5\ \text{J}^{-1}\text{cm}^2$.

Dependence of the signal amplitude on the spatial frequency of the interference pattern is another characteristic usually measured in non-steady-state photo-EMF experiments (Fig. 3.27). The observed behavior of the signal is explained as follows: the growth of the signal at low K is due to the increase of the space charge field amplitude, which is proportional to the diffusion field

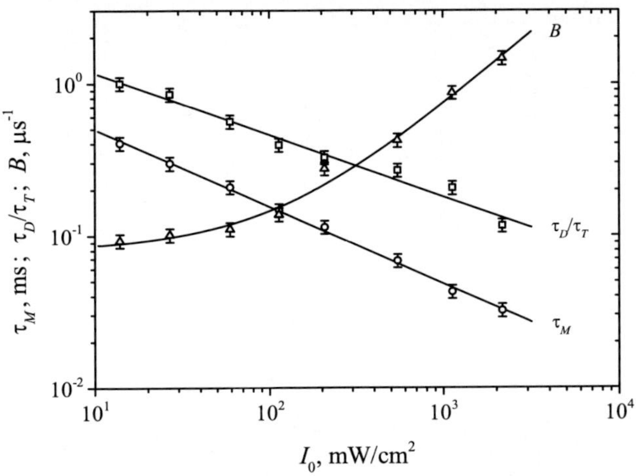

Figure 3.26. Dependencies of parameters τ_M, τ_D/τ_T and B versus light intensity. $K = 0.36\ \mu\text{m}^{-1}$.

E_D [2]. The decay at high K is caused by the diffusion "blurring" of the photoconductivity grating. The expression for this dependence is known from the classic theory of the non-steady-state photo-EMF effect [20, 38]:

$$j^{\omega}(K) \propto \frac{K}{1 + K^2 L_D^2}. \tag{3.57}$$

The maximum of the signal is achieved at $K = L_D^{-1}$, so the electron diffusion length can easily be estimated from the experimental curves: $L_D = 150 - 160$ nm. Figure 3.27 presents also the dependencies of the first cut-off frequency versus spatial frequency – experimental and calculated using the following expression with $L_D = 52 - 58$ nm:

$$\omega_1 = [\tau_M(1 + K^2 L_D^2)]^{-1}. \tag{3.58}$$

We should notice that increased light reflection at large incident angles leads to the decrease of light intensity in the crystal volume, and this can affect the signal excitation at $K > 20\ \mu\text{m}^{-1}$. The difference in the estimations of the diffusion length obtained from dependencies $J^{\omega}(K)$ and $\omega_1(K)$ may be associated with the mentioned feature of the experiment.

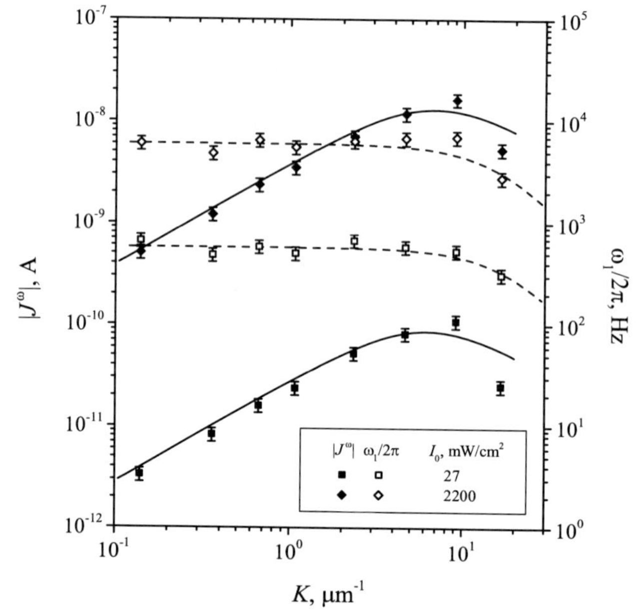

Figure 3.27. Dependencies of the signal amplitude and first cut-off frequency on the spatial frequency of the interference pattern. $I_0 = 27, 2200$ mW/cm^2.

An external high-frequency ac voltage can be utilized to increase the amplitude of the nonstationary holographic current [60, 155]. In this work we investigated the nonresonant signal enhancement which takes place when frequency of phase modulation is low compared to the frequency of an external field, i.e. $\omega \ll \Omega$ [60]. Frequency transfer functions of the holographic current excited in SiC at different effective values of ac field are presented in Fig. 3.28. The growth of the signal amplitude is resulted from the more efficient redistribution of the electrons between bright and dark interference fringes and from the corresponding larger amplitude of the space charge field grating [55]. The increase of the amplitude should be accompanied by the decrease of the cut-off frequency, but the frequency dependence should keep the shape typical for the diffusion

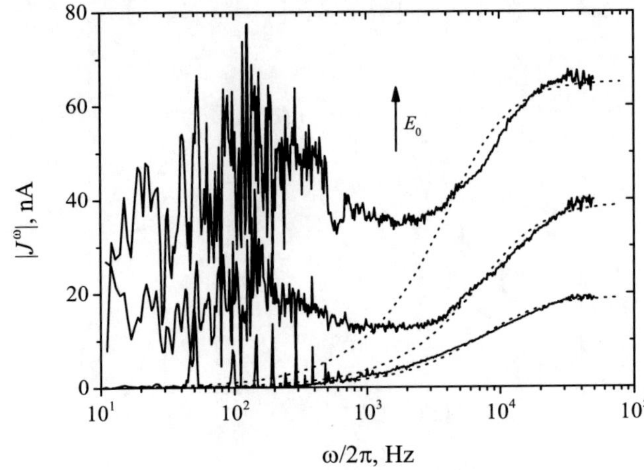

Figure 3.28. Frequency transfer functions of the nonstationary holographic current excited in external ac electric fields $E_0 = 0, 1.6, 3.6$ kV/cm. $I_0 = 2400$ mW/cm^2, $K = 2.6\ \mu$m^{-1}, $\Omega/2\pi = 100$ kHz.

regime of excitation [60]:

$$j^\omega = \frac{m^2\Delta}{2}\sigma_0 E_D \frac{-i\omega\tau_M(1+E_0^2/E_L^2)}{1+i\omega\tau_M(1+E_0^2/E_M^2)}, \tag{3.59}$$

$$\omega_1 = [\tau_M(1+E_0^2/E_M^2)]^{-1}. \tag{3.60}$$

Here E_0 is the effective value of external ac electric field, $E_L = k_BT/eL_D$ and $E_M = (K\mu\tau)^{-1}$ are characteristic values of this field, μ is the mobility of electrons. The behavior described by these formulae is observed in region $\omega/2\pi = 3 - 50$ kHz. Frequency dependencies in region $\omega/2\pi < 3$ kHz contain the maximum whose origin has no explanation in the frames of the existing theory.

Figure 3.29 presents the dependencies of the signal amplitude (measured at $\omega > \omega_1$) and cut-off frequency versus the effective value of applied ac field. Approximation of the experimental curves by Eqs. (3.59) and (3.60) provides estimation of the diffusion length: $L_D = 150 - 230$ nm.

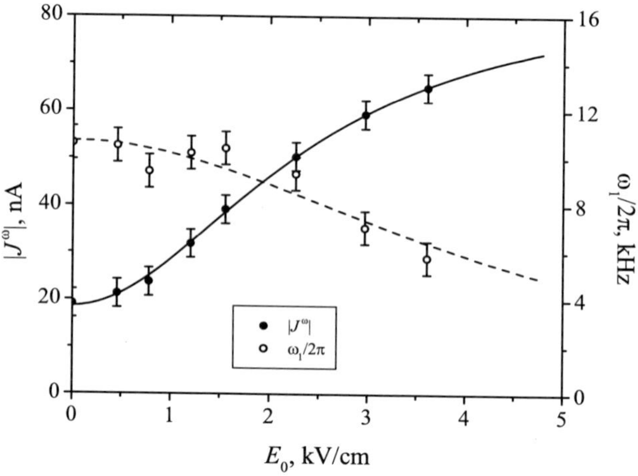

Figure 3.29. Dependencies of the signal amplitude and first cut-off frequency on the effective value of ac electric field. $I_0 = 2400$ mW/cm^2, $K = 2.6\ \mu$m^{-1}, $\Omega/2\pi = 100$ kHz.

Unexpected features of photoinduced effects have been observed in the neutron irradiated SiC crystal. Frequency dependencies of the photoconductivity response and nonstationary holographic current cannot be described in the frame of the simplest one-level model of semiconductor. The utilization of the two-level model partially eliminates this discrepancy. A more precise description of the experimental data can be achieved by the utilization of multilevel model of semiconductor with the several types of shallow traps. This requires the generalized theoretical approach, which can be considered as a task for the future. Meanwhile, the limitation of the frequency range and not very pronounced peculiarities in the transfer functions lead to the large spreading of the obtained parameters. For example, the values of conductivity estimated in the experiments with photoconductivity response and holographic current differ by more than order of magnitude.

The nonstationary holographic current in the SiC crystal placed in the ac electric field has also demonstrated the nontrivial behavior: there is a maximum at low frequencies instead of monotonous growth. Having compared Figs. 3.23

and 3.28 we can suppose that the appearance of this maximum is associated with the complicated relaxation of the photoconductivity in this frequency region. The heavy signal noisiness decaying after 1 kHz proves this assumption as well: spectral density of the generation-recombination noise decays starting from the frequency equal to inverse photoconductivity relaxation time [136].

Let us compare the obtained photoelectric parameters with the values reported earlier. The values of Maxwell relaxation time and $\mu\tau$-product had been estimated using the similar technique involving space-charge wave excitation in $6H$-SiC at the wavelength $\lambda = 488$ nm and light intensity $I_0 = 125$ mW/cm^2 [152]: $\tau_M = 0.52$ ms ($\sigma_0 = 1.6 \times 10^{-9}\ \Omega^{-1}\text{cm}^{-1}$) and $\mu\tau = 4.5 \times 10^{-7}$ cm^2/V ($L_D = 1.1\ \mu$m). Our crystal has 3.5 times higher photoconductivity ($\sigma_0 = 5.8 \times 10^{-9}\ \Omega^{-1}\text{cm}^{-1}$ for $I_0 = 110$ mW/cm^2) while its diffusion length is 5 – 20 times lower. The lower diffusion length in neutron irradiated crystal is an expected phenomenon, but the higher photoconductivity can hardly be explained: this contradiction may be originated from the different crystal growth procedures.

There is a lot of alternative methods, such as deep level transient spectroscopy, thermally stimulated current and Raman spectroscopy, which can provide information about the activation energy and concentration of traps in SiC crystals. The shallow traps associated with nitrogen atoms have activation energy of $E_T = 60 - 100$ meV and concentration of $M_T = 10^{11} - 10^{18}$ cm^{-3} [149, 158, 159]. The defects created by neutron radiation have activation energy of $E_T = 0.23 - 0.68$ eV and concentration of $M_T \sim 10^{14}$ cm^{-3} [160]. For example, choosing $E_T = 60$ meV and $M_T = 10^{17}$ cm^{-3} and the values for β and τ_T obtained in this paper we can roughly estimate the density of states in the conduction band [156]: $N_c = \beta\tau_T M_T \exp(E_T/k_B T) \sim 10^{19}$ cm^{-3}.

The results presented in the paper confirm the general type of changes in the electronic properties of SiC during radiation doping [149, 150]. The relatively high conductivity of initial non-irradiated crystals are usually attributed to the presence of nitrogen impurity producing shallow donor levels near the conduction band [149, 158, 159]. Irradiation by various high-energy particles, including reactor neutrons, creates a number of compensating acceptor centers, which increase the recombination rates and correspondingly decrease the electron lifetime and crystal conductivity.

In conclusion, we have demonstrated the capabilities of the nonstationary holographic currents in characterization of semiinsulating SiC crystal irradiated by reactor neutrons. The signal was studied in the absence and in the presence of an external ac electric field. In both cases the frequency transfer functions contain peculiarities, which are not described in the frames of the simplest one-level model of semiconductor. The utilization of the two-level model involving generation of electrons from the both deep centers and shallow traps noticeably improves the quantitative description of the effect. The photoelectric parameters such as photoconductivity, diffusion length of electrons, thermal excitation rate and ionization cross section for shallow traps are estimated. The further development of the technique can be associated with the generalization of theoretical approach to the multilevel model of semiconductor. The measurements of the temperature dependencies are also desirable: they may provide the activation energy of shallow traps. Moreover, the rather high amplitude of the signal in SiC makes this crystal a possible candidate for production of adaptive sensors of phase- and frequency-modulated optical signals [130, 135–142].

6. AlN

Aluminium nitride (AlN) stands out against other III-V compound semiconductors for its large band gap energy of 6.2 eV. This ensures its utilization in detection of visible and UV light, optical modulators, LEDs and lasers [161]. Chemical stability, high mechanical hardness (7 in Mohs scale) [162], high melting temperature (3070 K) [163], sufficient thermal conductivity (3.19 W cm^{-1} K^{-1}) [164] and electron mobility (300 cm^{2}V^{-1}s^{-1}) [165] also promote its application in electronic devices operating in difficult environments.

AlN crystal has wurzite hexagonal structure with unit-cell dimensions $a = 3.111$ Å and $c = 4.980$ Å [162]. Crystal lattice belongs to $6mm$ point symmetry group, which allows presence of the electrooptic effect in particular. The first, to our knowledge, observation of the photorefractive effect in AlN, i.e. photoinduced change of the refractive index by means of electrooptic effect, was reported in Ref. [166]. The photorefractive holograms recorded in this material demonstrated behavior and characteristics similar to those of "classical" photorefractive crystals ($Bi_{12}SiO_{20}$, $LiNbO_3$).

The photorefractive effect includes the stage of space charge formation, which occurs via migration of charge carriers, i.e. electric current [2]. This current is the basis of the non-steady-state photoelectromotive force (photo-EMF) effect, which reveals itself as an alternating electric signal arising in the sample illuminated by an oscillating interference pattern [20, 38]. Since the current is resulted from the periodic relative shifts of the photoconductivity and space charge gratings, the technique based on this effect allows determination of the number of photoelectric parameters (conductivity, carrier sign, lifetime, diffusion length, and drift mobility) and can be applied for the investigation of both non-centrosymmetric and centrosymmetric media. In this section we study the non-steady-state photo-EMF effect in AlN bulk crystal and apply this technique for determination the photoelectric properties of the material.

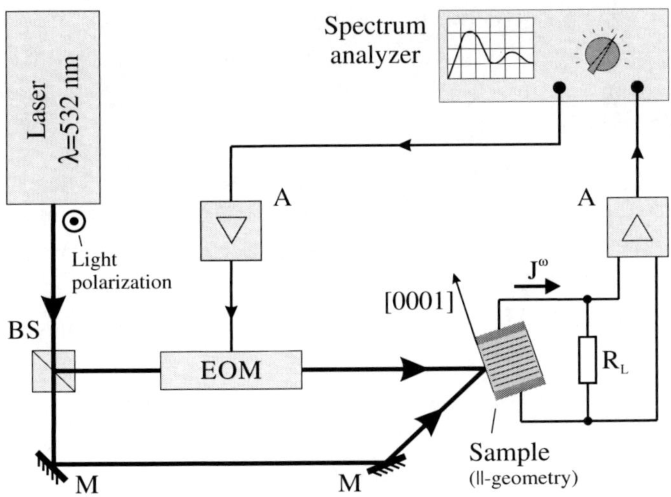

Figure 3.30. Experimental setup for investigation of the non-steady-state photo-EMF in semiconductors. *EOM* is the electrooptic modulator, *BS* is the beam splitter, *M* are the mirrors, *A* are the amplifiers. Axis [0001] is perpendicular to the figure plane for ⊥-geometry.

The experimental arrangement used for the investigation of the non-steady-state photo-EMF in AlN crystal is shown in Fig. 3.30. The second harmonic of Nd:YAG laser ($\lambda = 532$ nm, $P_{out} \simeq 20$ mW) laser was split into two beams forming the interference pattern with spatial frequency K and contrast $m = 0.48$ on the sample's surface. The signal beam was phase modulated with frequency ω and amplitude $\delta = 0.62$ by the electrooptic modulator ML-102A. The polarization plane of the light was perpendicular to the incidence plane. The beams were expanded in order to illuminate the whole interelectrode spacing. Two sample orientations were studied with the parallel orientation of the grating vector **K** and the optical c-axis [0001] in the former case ("longitudinal" $\|$-geometry), and with the perpendicular orientation in the latter one ("transversal" $\perp$-geometry). The photocurrent arising in the sample generated the corresponding voltage on the load resistor $R_L = 10$ MΩ. The voltage was amplified and measured using spectrum analyzer SK4-56 (detection bandwidth $\Delta f = 3$ Hz) or lock-in nanovoltmeter Unipan-232B (with integration time $\tau_{int} = 1 - 100$ s). The input resistance of the preamplifier was of 100MΩ. The sum of the sample and input capacitances was of 3 pF. The corresponding correction of the frequency transfer function was performed during calculation of the current amplitude.

The crystal was grown by PVT technique [167] and has pale yellow color. The characteristic dimensions of the sample were $6.0\times3.8\times3.0$ mm^3. The front and back surfaces (3.8×3.0 mm^2) were polished. The silver paste electrodes were painted on the lateral surfaces (6.0×3.0 mm^2 for $\|$-geometry and 6.0×3.8 mm^2 for $\perp$-geometry). The static dielectric constant of the material is of $\epsilon_{\|} = 9.14 - 9.32$ and $\epsilon_{\perp} = 7.76$ [168, 169].

The presence of the non-steady-state photo-EMF effect in AlN bulk crystal is the first result which should be pointed out. The amplitude of the detected signal is of order of $0.1 - 1$ pA which is noticeably lower than that in the model objects – sillenite crystals $Bi_{12}Si(Ti,Ge)O_{20}$, where it reaches 1 nA [20]. Nevertheless this amplitude is quite enough for consistent detection of the signal with signal-to-noise ratio of $1 - 10$. The phase of the detected signal indicates that electrons prevail in the process of photo-EMF excitation at low and moderate spatial frequencies of the interference pattern.

We have measured the frequency transfer functions of the non-steady-state photo-EMF signal (Fig. 3.31) in the "longitudinal" ($\|$) and "transversal" ($\perp$) geometries of the experiments, i.e. for the cases when direction of the space charge field $\mathbf{E}_{SC}$, grating vector **K**, and photocurrent $\mathbf{J}^{\omega}$ is parallel or perpendicular to the c-axis. The signal demonstrates typical behavior, namely, there

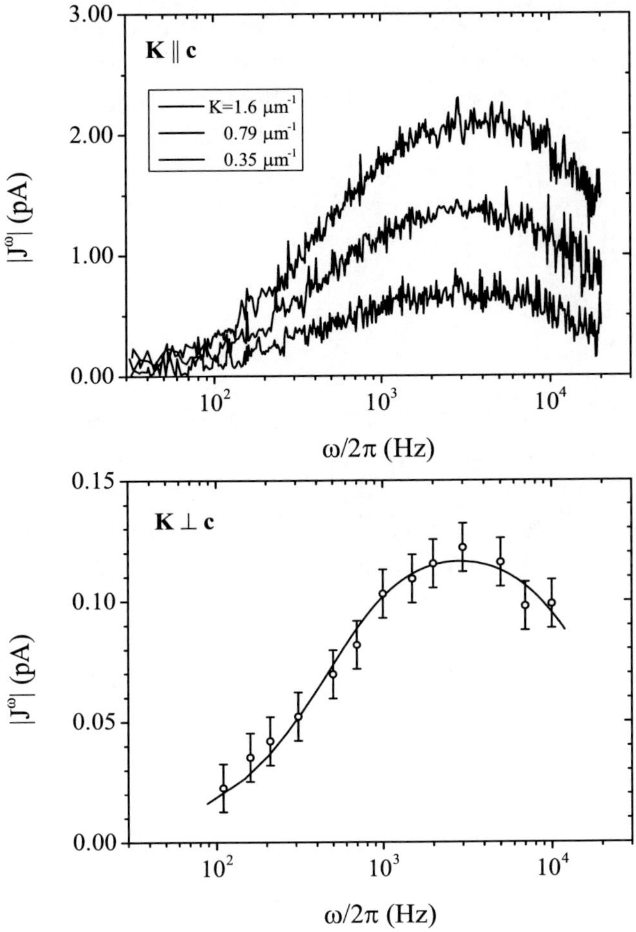

Figure 3.31. Frequency transfer functions of the non-steady-state photo-EMF signal in AlN crystal. The dependencies are measured for two geometries: **K** $\parallel$ c-axis, $K = 0.35\ \mu\text{m}^{-1}$, $0.79\ \mu\text{m}^{-1}$, $1.6\ \mu\text{m}^{-1}$ (top figure), and **K** $\perp$ c-axis, $K = 2.0\ \mu\text{m}^{-1}$ (bottom figure). $I_0 = 110$ mW/cm^2. The solid line shows approximation by Eq. (3.61) with $\omega_1/2\pi = 600$ Hz and $\omega_2/2\pi = 14$ kHz.

is a linear growth of the amplitude for low frequencies of the phase modulation $\omega < \omega_1$, frequency independent region for higher frequencies $\omega_1 < \omega < \omega_2$, and decay for frequencies $\omega > \omega_2$. These regions are separated by the so-called cut-off frequencies $\omega_1/2\pi = 140 - 700$ Hz and $\omega_2/2\pi = 13 - 30$ kHz. These dependencies are well described by the following expression which is equivalent to that obtained earlier for the simplest model of the semiconductor crystal with one type of partially compensated donor centers [38]:

$$J^\omega = J_m^\omega \frac{i\omega(\omega_2 - \omega_1)}{\omega_1\omega_2 - \omega^2 + i\omega(\omega_2 - \omega_1)}, \tag{3.61}$$

here J_m^ω is the current amplitude at the maximum of frequency dependence. The linear growth for low modulation frequencies ($\omega/2\pi < 700$ Hz in Fig. 3.31) is an important manifestation of the adaptive nature of space charge formation in wide-gap semiconductor materials. The signal in this frequency region is small since both the space charge field grating and grating of free carriers (photoconductivity grating) follow the movement of the interference pattern. The spatial shift between the gratings is nearly equal to $\pi/2$ which results in low signal amplitude. For higher modulation frequencies (frequency independent region) the grating with larger relaxation time can be considered as "frozen in", the periodic spatial shifts increase, and the photo-EMF signal reaches its maximum. For very high modulation frequencies ($\omega/2\pi > 13$ kHz in Fig. 3.31) the oscillation amplitudes of both gratings as well as corresponding relative shifts between them become negligible, and the resulting current vanishes.

The dependence of the signal amplitude versus spatial frequency of the interference pattern, i.e. $J^\omega(K)$, is another important characteristic, that is usually measured in the non-steady-state photo-EMF experiments (Fig. 3.32). The behavior of the signal can be easily explained: the signal increases for low K due to the growth of space charge field amplitude which is proportional to the so-called diffusion field $E_D = Kk_BT/e$, the signal decrease for high K is resulted from the diffusion blurring of the conductivity grating. The dependencies were fitted by the following simple expression known from the non-steady-state photo-EMF theory [20]:

$$J^\omega(K) \propto \frac{\sigma_0 E_D}{1 + (KL_D)^2}. \tag{3.62}$$

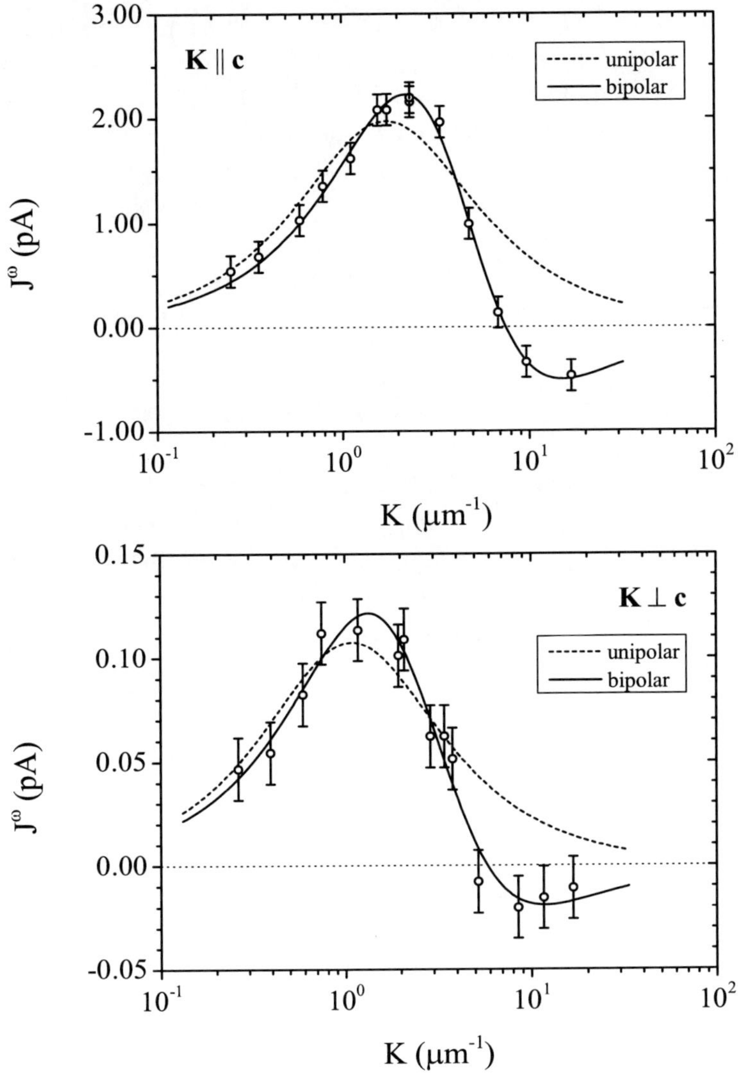

Figure 3.32. Dependence of the non-steady-state photo-EMF amplitude on the spatial frequency of the interference pattern. The dependencies are measured for two geometries: $\mathbf{K} \parallel c$-axis (top figure), and $\mathbf{K} \perp c$-axis (bottom figure). $I_0 = 110$ mW/cm^2. Dotted and solid lines show approximation by Eqs. (3.62) and (3.63).

The diffusion length of electrons was estimated from the measured dependence: $L_{D\|} = 0.58$ μm and $L_{D\perp} = 0.92$ μm for the $\|$- and $\perp$-geometries, respectively. As seen the signal amplitude decays faster than K^{-1} and changes its sign for high spatial frequencies. A similar effect was observed earlier in GaAs crystal at $\lambda = 1.15$ μm [86]. Such behavior is explained by the presence of bipolar photoconductivity: the sign of the signal for low spatial frequencies is determined by the carriers with larger photoconductivity, while the carriers with larger generation rate define the sign of the signal for high K. The expression for photo-EMF amplitude can be written in this case as follows:

$$J^{\omega}(K) \propto \frac{\sigma_0^e E_D}{1+(KL_D^e)^2} - \frac{\sigma_0^h E_D}{1+(KL_D^h)^2}, \tag{3.63}$$

where $\sigma_0^{e,h}$ and $L_D^{e,h}$ is the average photoconductivities and diffusion lengths of electrons and holes, respectively. One can note that this expression fits experimental dependencies much better (Fig. 3.32), and it describes the change of the signal sign. The best fit was obtained for the following combination of material parameters: $\sigma_{0\|}^e/\sigma_{0\|}^h = 1.8$, $L_{D\|}^e = 0.27$ μm, $L_{D\|}^h = 0.18$ μm for $\|$-geometry and $\sigma_{0\perp}^e/\sigma_{0\perp}^h = 5.3$, $L_{D\perp}^e = 0.58$ μm, $L_{D\perp}^h = 0.20$ μm for $\perp$-geometry.

Dependencies $\omega_1(K)$ and $\omega_2(K)$ measured in $\|$-geometry are presented in Fig. 3.33. The expressions for the cut-off frequencies obtained earlier for the simplest model of semiconductor with one type of compensated donor centers are as follows [38]:

$$\omega_1 \simeq [\tau + \tau_M(1+K^2L_D^2)]^{-1}, \tag{3.64}$$

$$\omega_2 = \omega_1 + \tau_M^{-1} + (1+K^2L_D^2)\tau^{-1}. \tag{3.65}$$

The following parameters were determined from the experimental dependencies: $\tau_{M\|} = 0.23$ ms, $\tau_{\|} = 11$ μs, $L_{D\|} = 0.17$ μm. Using the measured value of Maxwell relaxation time $\tau_M = \epsilon\epsilon_0/\sigma_0$ one can easily estimate the average conductivity σ_0 of the material. For the sample in $\|$-geometry it equals: $\sigma_{0\|} = 3.6 \times 10^{-9}$ Ω^{-1}cm^{-1}.

Characteristic relaxation times $\tau_{M\perp}$ and $\tau_{\perp}$ for $\perp$-geometry were also estimated from the only transfer function shown in Fig. 3.31: $\tau_{M\perp} \simeq 0.26$ ms, $\tau_{\perp} \simeq 12$ μs. The corresponding photoconductivity equals $\sigma_{0\perp} \simeq 2.6 \times 10^{-9}$ Ω^{-1}cm^{-1}.

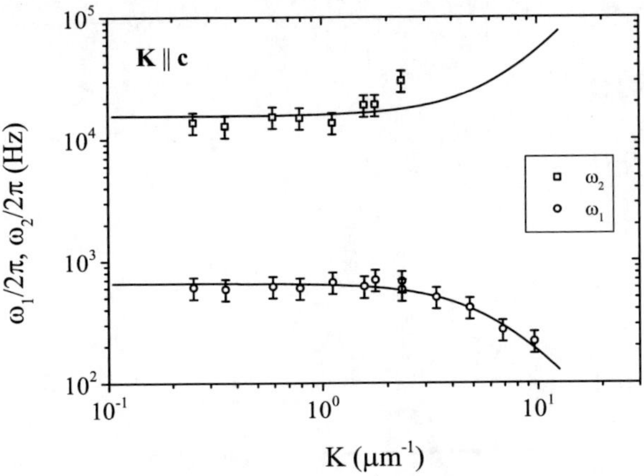

Figure 3.33. Dependencies of the first ω_1 and second ω_2 cut-off frequencies on the spatial frequency of the interference pattern. $I_0 = 110$ mW/cm^2. Solid lines show the approximation by Eqs. (3.64) and (3.65).

Dependence of the non-steady-state photo-EMF versus average light intensity was studied as well (Fig. 3.34). Nonlinear growth of the first cut-off frequency and signal amplitude in the maximum of transfer function was observed: $\omega_1 \propto I_0^{0.67}$ and $J^\omega \propto I_0^{0.81}$. These results correlate with the nonlinearity of holographic recording in AlN [166], although the powers of function I^x are different. The second cut-off frequency remains nearly constant, what means the independence of the conductivity relaxation time on light intensity.

The experiments on non-steady-state photo-EMF excitation have revealed strong anisotropy of the AlN crystal: the photocurrent density along c-axis is about 25 higher than the one across c-axis. In the frames of unipolar conductivity this result can be attributed solely to anisotropy of the photoconductivity, since $j^\omega \propto \sigma_0 E_D$, while diffusion field $E_D \propto K$ remains comparable for the both orientations. In the frames of bipolar conductivity the photo-EMF amplitude is the difference of the electron and hole parts: $j^\omega \propto (\sigma_0^e C^e - \sigma_0^h C^h) E_D$, where $C^{e,h}$ are the coefficients taking into account the diffusion blurring of conductivity grating [see Eq. (3.63)]. Thus the discussed anisotropy can be attributed to both the absolute values σ_0^e, σ_0^h and their relation σ_0^e/σ_0^h.

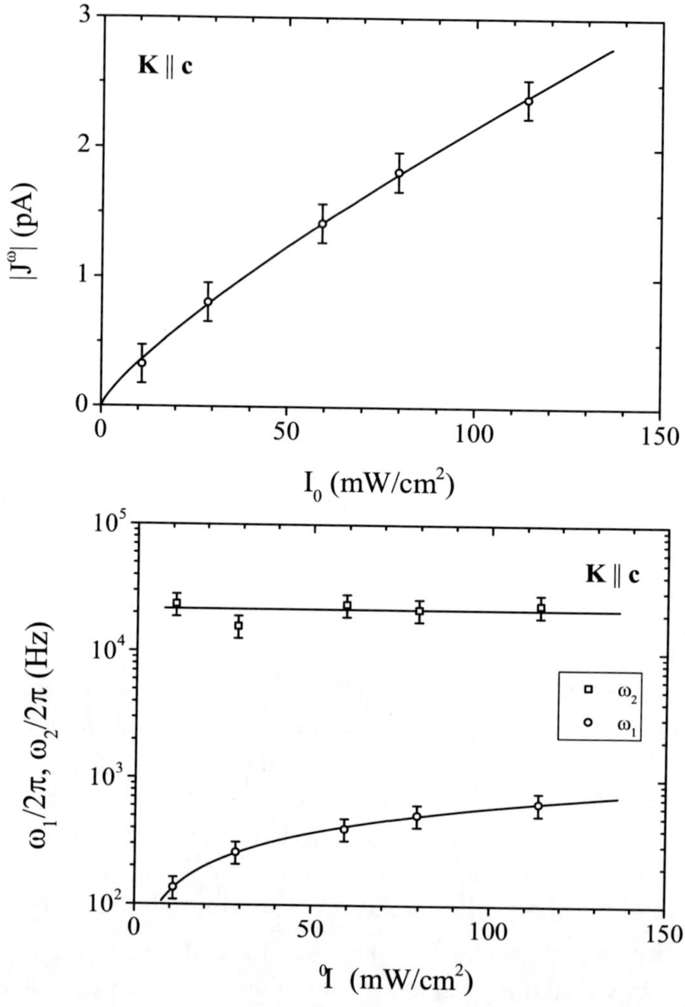

Figure 3.34. Dependencies of the non-steady-state photo-EMF amplitude (top figure) and cut-off frequencies (bottom figure) versus average light intensity. $\mathbf{K} \parallel c$-axis, $K = 1.9\ \mu\mathrm{m}^{-1}$.

There is a serious contradiction in the experimental data and theoretical approach used. As it was mentioned above, the current along c-axis is higher that that in transverse direction (Fig. 3.32), whereas dependence $J_\perp^\omega(K)$ seems to be shifted to lower spatial frequencies with respect to dependence $J_\parallel^\omega(K)$. Intuitively the opposite shift is expected, since the larger current amplitude in direction of c-axis implies the larger photoconductivities, larger $\mu\tau$-products, and larger diffusion lengths. Quantitatively we can describe this contradiction as follows. The photocurrent amplitude for low K can be written as $j^\omega(K) \propto (\sigma_0^e - \sigma_0^h)E_D$. Then the ratio of signals along and across c-axis is represented as

$$\frac{j_\parallel^\omega}{j_\perp^\omega} = \left(\frac{L_{D\parallel}^h}{L_{D\perp}^h}\right)^2 \frac{\sigma_{0\parallel}^e/\sigma_{0\parallel}^h - 1}{\sigma_{0\perp}^e/\sigma_{0\perp}^h - 1}. \tag{3.66}$$

Here we assumed the generation rates of electron and holes to be independent on light polarization. After substitution of material parameters estimated from Fig. 3.32 we obtain $j_\parallel^\omega/j_\perp^\omega = 0.15$, i.e. the calculated photocurrent along c-axis is not higher but lower than that across c-axis. Now we can not explain this contradiction, we can just speculate with possible polarization dependence of generation rates mentioned above. The relative errors in determination of material parameters are of the order of 10% and resulted mainly from the instability of the laser's output power.

The previous studies [162, 170, 171] of charge transport in undoped AlN provided rather wide range for the dark conductivity: $\sigma_d = 10^{-12} - 10^{-9}$ $\Omega^{-1}\text{cm}^{-1}$. In this work we obtained the conductivity values of the order of 10^{-9} $\Omega^{-1}\text{cm}^{-1}$ as well. Since the noticeable growth of the first cut-off frequency versus light intensity was revealed (Fig. 3.34), we can state that photoconductivity exceeds the dark conductivity, and space charge field grating is recorded effectively for the used illumination wavelength and power level. We can also roughly estimate the mobility of carriers using the obtained values of diffusion lengths L_D and lifetimes τ. For electrons, which dominate in photo-EMF excitation, we obtain $\mu_\parallel^e \sim 3 \times 10^{-3}$ $\text{cm}^2\text{V}^{-1}\text{s}^{-1}$ and $\mu_\perp^e \sim 1 \times 10^{-2}$ $\text{cm}^2\text{V}^{-1}\text{s}^{-1}$. These values are sufficiently smaller than those calculated (300 $\text{cm}^2\text{V}^{-1}\text{s}^{-1}$) and measured (125 $\text{cm}^2\text{V}^{-1}\text{s}^{-1}$) earlier [165, 172]. Such difference is often attributed to the trapping processes, which can reduce the average velocity of charge migration (and corresponding effective mobility) by several orders.

Nonlinear photoconductivity ($\sigma_0 \propto \omega_1 \propto I_0^{0.67}$) along with intensity independent relaxation time ($\tau \propto \omega_2 = \mathrm{const}(I_0)$) give rise some contradiction in the frames of simplest model of unipolar semiconductor: nonlinear photoconductivity is usually accompanied by the change of relaxation time. In the case of bipolar semiconductor the relative contributions of electrons and holes to the overall conductivity and its relaxation time can change with the light intensity and this discrepancy may not be significant.

Generally speaking, the bipolar conductivity of AlN crystal is unexpected in itself. The energy of photons is of 2.3 eV, whereas the band gap of the material is of 6.2 eV. This means that electrons and holes can not be generated from the same local levels. There must be at least two groups of levels in the forbidden gap providing extrinsic absorption of the green light and corresponding independent generation of electrons and holes.

To summarize we studied the effect of non-steady-state photo-EMF in new photosensitive material – aluminium nitride bulk crystal. The significant anisotropy of the material was revealed in the experiments with excitation of the photo-EMF along and across [0001] axis. The presence of the bipolar photoconductivity was established from the change of the signal sign for high spatial frequencies. The average photoconductivity of the material was estimated from the measurements of the signal amplitude versus frequency of phase modulation. The dependence of the signal versus spatial frequency provided the estimation of the diffusion lengths of electrons and holes.

The obtained results have demonstrated that the non-steady-state photo-EMF technique can be considered as a powerful tool for investigation of wide band gap semiconductors. A number of photoelectric parameters are estimated using the only arrangement. In contrast to the standard methods of semiconductor characterization the technological problem of ohmic contacts for high-resistive materials is not so significant here, since the effect of the non-steady-state photo-EMF is based on the excitation of an alternating current.

7. Ga_2O_3

The investigation of transport properties in high-resistive semiconductors is often complicated not only by the small values of electric current, diffusion and drift lengths but also by the formation of the space charge near blocking contact. An elegant solution of this problem was proposed, when the non-steady-state photo-EMF effect, or nonstationary holographic current in other terminology, was discovered [19, 20, 38]. The effect reveals itself as an alternating electric current arising in a semiconductor illuminated by an oscillating interference pattern. The related techniques realized with an interference pattern running with constant velocity are called as holographic current [131] or moving photocarrier grating technique [32]. The space charge formation produces no harmful action on these effects; on the contrary, it takes part in the excitation of the signal, because the resulting electric current appears due to spatial shifts of the space-charge and photoconductivity distributions in the crystal volume.

The excitation of the non-steady-state photo-EMF signal involves processes of carrier generation, diffusion, drift in an electric field and recombination to local centers. This complex nature of the effect makes it a powerful tool for determination of photoelectric parameters such as type and value of photoconductivity, carrier lifetime, mobility and diffusion length, density of recombination centers, etc. [32, 38, 132–134, 173, 174]. Besides the tasks of material characterization the non-steady-state photo-EMF is also used for the detection of phase- and frequency-modulated optical signals [21, 136, 138, 139, 141]. Being defined by the drift component of the current averaged over interelectrode spacing the non-steady-state photo-EMF is an integral effect. This feature as well as the fact, that arising current is alternating, allow observation of the effect not only in single crystals but also in polycrystalline, amorphous, composite and polymer media. Moreover, these also allow the detection of optical signals with complex wavefronts, even speckle light patterns.

In this section we apply the non-steady-state photo-EMF technique for the investigations of a β-Ga_2O_3 crystal [175]. The interesting optical and electrical properties of this material attracts the attention of researchers during last years. The band gap of β-Ga_2O_3 is of 4.8 eV, so the crystal is highly transparent for the visible light and partially for UV. This advances the development of solar-blind detectors of deep ultraviolet radiation based on nanowires [176], nanobelts [177], thin films [178, 179], and bulk crystals [180]. Oxygen vacancies endow a β-Ga_2O_3 crystal with the electronic conductivity, and this can be utilized for the fabrication of transparent electrodes [181]. The refractive index $n \simeq 1.9$ can make such electrodes deposited on A3B5 semiconductors to be antireflective for the certain wavelength, e.g., 885 nm for Ga_2O_3/GaAs structure [182]. Thin films of Ga_2O_3 grown with oxygen deficiency exhibit the switching memristor behavior promising for nonvolatile memory applications [183]. The crystal possesses the photocatalytic properties and can be used as an photoelectrode for the water splitting under UV radiation [184]. The application of the material for the development of field-effect transistors and gas sensors was also proposed [185–187]. In spite of the great interest to β-Ga_2O_3 for ultraviolet applications, the perspective of the crystal utilization in the visible and even infrared range should not be ignored. Here we report on the photoelectric properties at $\lambda = 532$ nm ($h\nu = 2.33$ eV) and reveal the material's anisotropy measuring the signal along the [100] and [010] axes. Both the non-resonant and resonant techniques of the photo-EMF excitation are implemented in the experiments.

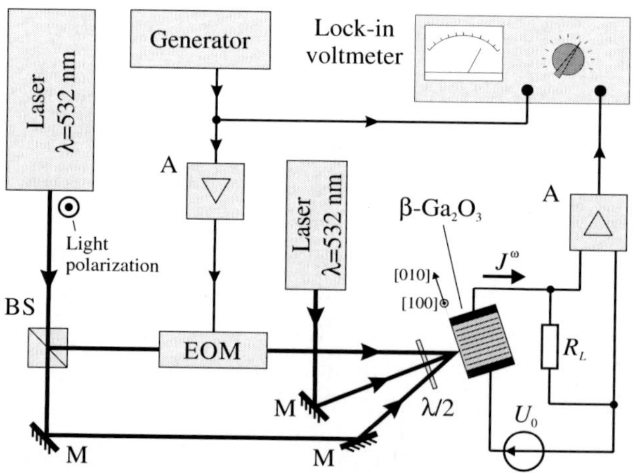

Figure 3.35. Experimental setup for the investigation of the non-steady-state photo-EMF. EOM is the electrooptic modulator, BS is the beam splitter, M is the mirror, A is the amplifier.

The experiments with the excitation of the non-steady-state photo-EMF in β-Ga_2O_3 are carried out with the arrangement (Fig. 2.1) used earlier for investigations of other wide-bandgap semiconductors and nanostructured materials [173,174]. The second harmonic of Nd:YAG dc laser with the wavelength of $\lambda = 532$ nm is split into two beams, which then create the interference pattern with spatial frequency K, contrast $m = 0.98$ and average intensity I_0 on the crystal surface. The electro-optic modulator introduces phase modulation with amplitude $\delta = 0.61$ and frequency ω into the signal beam. The photocurrent arising in the sample produces a voltage across the load resistor, which is amplified and then measured by the lock-in voltmeter. The experiments with an applied external voltage are performed at higher light intensities and lower contrast $m = 0.11$. This is achieved by additional illumination using the powerful laser pointer operating at the same wavelength. The polarization plane containing electric field vector is perpendicular to the incidence plane (TE-polarization) in the most of the experiments. The placement of the half-wave plate in front of the sample allows the rotation of the polarization plane, when it is necessary.

β-Ga_2O_3 is a monoclinic crystal with the cell dimensions $a = 12.23$ Å, $b = 3.04$ Å, $c = 5.80$ Å and $\beta = 103.7°$ [188]. The band gap is indirect with $E_g = 4.84$ eV [189]. The static dielectric constant of the material is a tensor with eigenvalues $\epsilon_{11} = 10.84$, $\epsilon_{22} = 11.49$ and $\epsilon_{33} = 13.89$ [190].

The crystal of gallium oxide was grown by the floating zone method at URN-2-ZM machine produced in MPEI [191]. The single crystal was grown from the ceramic bar obtained by the conventional technology. The seed was oriented so that the crystal grew along the [010] direction. The growth rate of 10 mm/h was chosen to ensure the stability of the crystallization processes. The quality of the obtained single crystals depends on the type of the atmosphere and its pressure. The studied crystal was grown in the 60 bar oxygen atmosphere, which provided the highest sample's quality.

The investigation of the crystal using conventional X-ray diffraction and light transmission spectroscopy was performed in the previous work [192]. The X-ray Laue method allowed us to orient the crystal, confirmed the lattice constants known from literature [188], and showed the absence of the crystal twinning. The light transmission spectrum practically coincides with the previously reported spectra [193–195], i.e., the crystal demonstrates the transmittance about 80% in the visible and near infrared ranges $400 - 1100$ nm and the absorption edge below 270 nm corresponding to interband transitions. There is also a shoulder at $270 - 275$ nm, which can be due to both the crystal structure and the presence of Ga^{3+} vacancies [195].

The sample used in our studies has the dimensions of $2.00 \times 2.15 \times 1.35$ mm along the crystallographic directions [100], [010] and direction perpendicular to the plane (001), respectively. The front and back surfaces (2.00×2.15 mm) are the (001) crystal's cleaved facets, no additional treatment was applied to them. The silver paste electrodes are deposited on the opposite pairs of lateral surfaces in the experiments with the light pattern grating vector $\mathbf{K}||[100]$ and $\mathbf{K}||[010]$.

The typical signal amplitude is of $10^{-12} - 10^{-9}$ A, which is two orders of magnitude lower than the values observed in model sillenite crystals $Bi_{12}Si(Ti,Ge)O_{20}$ [20, 155]. Nevertheless such an amplitude is sufficient for the detection with the signal-to-noise ratio of $0 - 60$ dB. The phase of the signal clearly points to the electron type of photoconductivity.

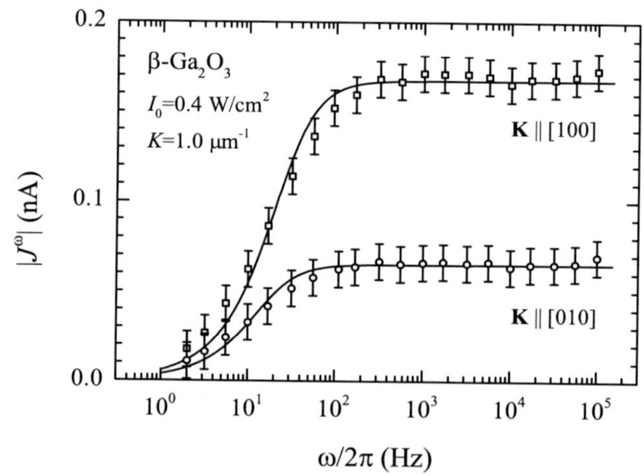

Figure 3.36. Frequency transfer functions of the non-steady-state photo-EMF J^{ω} measured for two crystal orientations.

We have measured the frequency transfer functions of the non-steady-state photo-EMF amplitude in the as fabricated β-Ga_2O_3 crystal (Fig. 3.36). The dependencies contain the growth of the signal at low frequencies ω followed by the frequency independent part. The signal growth at low frequencies is an important manifestation of the adaptivity of space charge formation in photoconductive materials [2, 20]. The signal at low ω is small because of the fact, that both the space charge field grating and the grating of free electrons (photoconductivity grating) follow the movement of the interference pattern. The spatial shift between them is nearly equal to $\pi/2$, which finally results in small value of the average drift component of the current. The grating with larger relaxation time becomes "frozen-in" at higher frequencies, the spatial shifts between gratings increase, and holographic current reaches its maximum at the frequency-independent region. Since there is no decaying part in the dependence, the space charge formation occurs in the condition of quasi-stationary photoconductivity ($\omega\tau \ll 1$), and the signal behavior is described by an expression [20]

$$j^{\omega} = \frac{m^2\Delta}{2}\sigma_0 E_D \frac{-i\omega\tau_M}{1 + i\omega\tau_M(1 + K^2 L_D^2)}. \qquad (3.67)$$

Here $E_D = (k_B T/e)K$ is the diffusion field, $\tau_M = \epsilon_0\epsilon/\sigma_0$ is the Maxwell relaxation time, L_D is the diffusion length of electrons [2]. The growing and frequency-independent regions are separated by the cut-off frequency ω_1

$$\omega_1 = [\tau_M(1 + K^2 L_D^2)]^{-1}. \qquad (3.68)$$

The dependencies of the signal amplitude and the first cut-off frequency versus illumination level are presented in Fig. 3.37. The amplitude of the photocurrent in the maximum of frequency transfer function is proportional to the light intensity: $J^{\omega} \propto I_0$. There is, however, a sublinear dependence of the cut-off frequency $\omega_1 \propto I_0^{0.69}$ for $\mathbf{K}||[100]$ and $\omega_1 \propto I_0^{0.51}$ for $\mathbf{K}||[010]$. This can be due to the corresponding nonlinearity of the photoconductivity. The presence of dark conductivity can contribute in the initial stage of dependence $\omega_1(I_0)$ as well. As seen from Eq. (3.68) the measurements of the cut-off frequency ω_1 at low K provide the estimate of the Maxwell relaxation time and corresponding photoconductivity of the sample. For the chosen intensities $I_0 = 0.039 - 0.40$ W/cm^2 of TE-polarized light the photoconductivity equals $\sigma_0 = (0.48 - 2.5) \times 10^{-10}$ Ω^{-1}cm^{-1} along the [100] axis and $\sigma_0 = (0.32 - 1.0) \times 10^{-10}$ Ω^{-1}cm^{-1} along the [010] axis.

Dependence of the signal amplitude on the spatial frequency of the interference pattern is another characteristic usually measured in the non-steady-state photo-EMF experiments (Fig. 3.38). The observed behavior of the signal is explained as follows: the growth of the signal at low K is due to the increase of the space charge field amplitude, which is proportional to the diffusion field

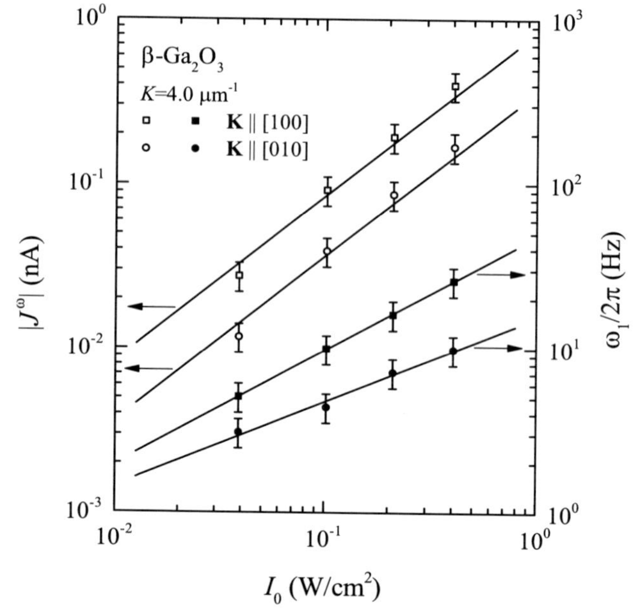

Figure 3.37. Dependencies of the signal amplitude J^{ω} and first cut-off frequency ω_1 on the average light intensity I_0.

E_D [2]. The decay at high K is caused by the diffusion "blurring" of the photoconductivity grating. The expression for this dependence is known from the theory of the non-steady-state photo-EMF effect [20, 38]:

$$j^{\omega}(K) \propto \frac{K}{1 + K^2 L_D^2}. \tag{3.69}$$

The maximum of the signal is achieved at $K = L_D^{-1}$, so the electron diffusion length can easily be estimated from the experimental curves: $L_D = 190$ nm for $\mathbf{K}||[100]$ and $L_D = 200$ nm for $\mathbf{K}||[010]$. Figure 3.38 also presents the dependencies of the first cut-off frequency versus spatial frequency. These experimental dependencies are fitted using Eq. (3.64) with $L_D = 110$ nm for $\mathbf{K}||[100]$ and $L_D = 130$ nm for $\mathbf{K}||[010]$. We should note that increased light reflection at large incident angles leads to the decrease of the light intensity in

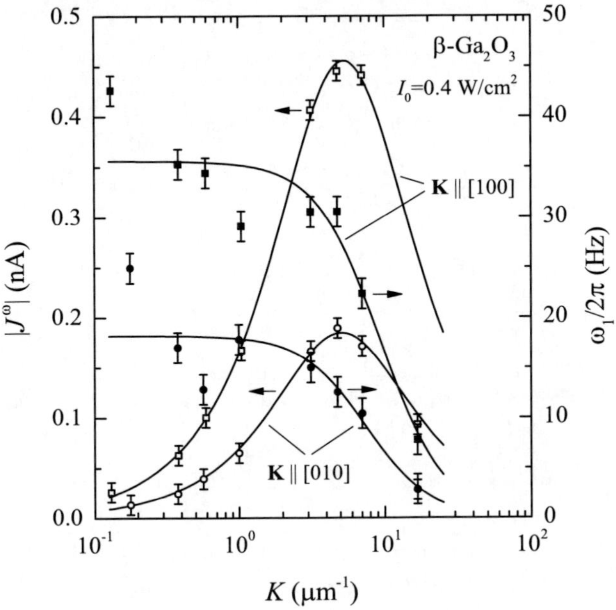

Figure 3.38. Dependencies of the signal amplitude J^ω and first cut-off frequency ω_1 on the spatial frequency K of the interference pattern.

the crystal volume, and this can affect the signal excitation for $K > 10\ \mu\text{m}^{-1}$. The difference in the estimates of the diffusion length obtained from dependencies $J^\omega(K)$ and $\omega_1(K)$ may be associated with the mentioned factor.

Looking at Figs. 3.36-3.38 one can easily note the anisotropy of the non-steady-state photo-EMF in the monoclinic β-Ga_2O_3. The fact that the photo-EMF amplitude and cut-off frequency in these figures are larger for the case $\mathbf{K}, \mathbf{j}||[100]$ does not make this geometry more effective for the signal excitation. When analyzing the anisotropy of the effect it is necessary to consider the dependence of the signal amplitude on the light polarization as well (Fig. 3.39). Each of the two curves plotted in Fig. 3.39 reveals the anisotropy of the photoexcitation processes, while the difference of the maximal (minimal) values points out to the anisotropy of the transport properties along the [100] and [010] directions. As seen from the figure, the maximal amplitudes are achieved in the

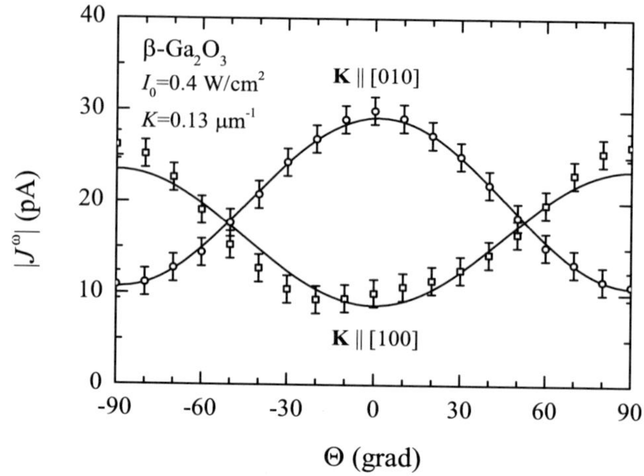

Figure 3.39. Dependencies of the non-steady-state photo-EMF amplitude J^{ω} on the angle between the incidence and polarization planes.

both orientations, when the light is polarized along the [010] axis. The maxima for the two geometries are observed at angles that differ by 90°, but their magnitudes are nearly equal, which points out to the weak anisotropy of the transport characteristics, e.g., the mobility of electrons. The signal in the case $\mathbf{K}, \mathbf{j}||[010]$ and $\Theta = 0°$ is even higher by $\sim 10\%$ than that for $\mathbf{K}, \mathbf{j}||[100]$ and $\Theta = 90°$. So the optimal geometry for the excitation of the non-steady-state photo-EMF in a β-Ga_2O_3 crystal is the following: the grating vector $\mathbf{K}$, holographic current $\mathbf{j}$ and light polarization should be parallel to the [010] axis.

In order to find out the origin of the polarization dependence (photoexcitation anisotropy) we have measured the sample's transmittance for the laser radiation and estimated the absorption coefficient for the two polarization states, i.e., along the [100] and [010] axes. The obtained values $\alpha_{[100]} \simeq 1.1$ cm^{-1} and $\alpha_{[010]} \simeq 1.3$ cm^{-1} differ by only 20%, while the maxima in Fig. 3.39 exceed corresponding minima by 280%. It should be noted here, that we have taken into account the reflection from the two crystal's interfaces for the both polarization states with $n_{[100]} = 1.9201$ and $n_{[010]} = 1.9523$ [196]. The calculated absolute values of α are still very rough estimates, since the sample's surfaces are not polished, and the volume has some optical inhomogeneity. So the polarization dependence of the absorption coefficient is not the primary source of the photoexcitation anisotropy. There is another factor that defines the generation rates of the electrons, namely, the quantum efficiency of the photoconductivity. The anisotropy of this parameter $\beta_{[100]}/\beta_{[010]} \simeq 2.3$ originates the observed polarization dependence of the detected signal.

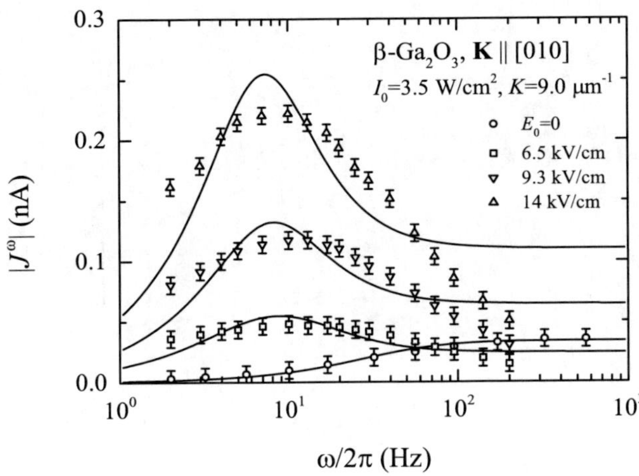

Figure 3.40. Frequency transfer functions of the nonstationary holographic current J^{ω} excited in an external dc electric field.

An external dc voltage can be utilized for increasing the amplitude of the non-steady-state photo-EMF [38,61,74,151]. The signal enhancement is not the only change arising in the dc field, the frequency dependence becomes resonant (Fig. 3.40). The appearance of the resonant maximum is due to the excitation of space-charge waves, which are the eigenmodes of space-charge evolution in semiconductors placed in an external dc field [2]. The theory of the effect provides the following expression for the signal amplitude [61]:

$$
\begin{aligned}
j^{\omega} &= \frac{m^2\Delta}{4}\sigma_0\left[\frac{i2E_0-\omega\tau_M(E_0+iE_D)}{1+i\omega\tau_M(1+K^2L_D^2+iKL_0)}\right. \\
&\quad \left. -\frac{i2E_0-\omega\tau_M(E_0-iE_D)}{1+i\omega\tau_M(1+K^2L_D^2-iKL_0)}\right],
\end{aligned}
\tag{3.70}
$$

where $L_0 = \mu\tau E_0$ is the drift length of electrons in the dc electric field E_0. The approximation of the experimental curves in Fig. 3.40 allows us to estimate the $\mu\tau$-product of electrons $\mu\tau \simeq 5.3 \times 10^{-10}$ cm^2/V along the [010] axis and the Maxwell relaxation time $\tau_M \simeq 16$ ms, which corresponds to the average photoconductivity $\sigma_0 \simeq 1.0 \times 10^{-10}\ \Omega^{-1}\text{cm}^{-1}$ ($I_0 = 3.5$ W/cm^2).

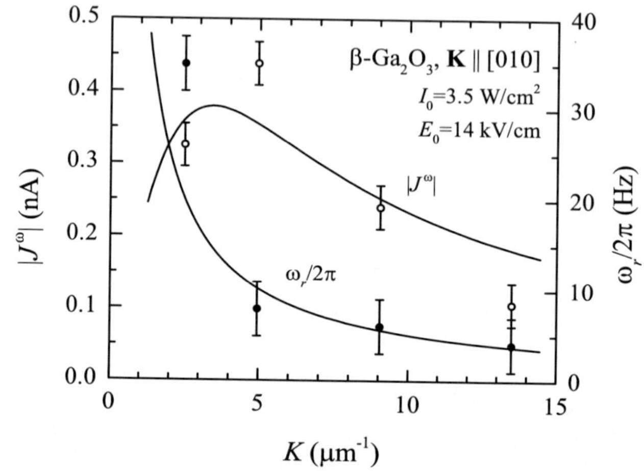

Figure 3.41. Dependencies of the resonant signal amplitude $J^\omega(\omega_r)$ and resonant frequency ω_r on the spatial frequency K of the interference pattern.

The space-charge waves run along the applied electric field and have an unusual dispersive law, i.e., dependence $\omega_{scw}(K)$:

$$\omega_r = \omega_{scw} = [\tau_M K L_0]^{-1}. \tag{3.71}$$

Such a decrease of the resonant frequency is observed in the experiment (Fig. 3.41). The resonant amplitude of the signal depends on the spatial frequency as well. It was shown earlier [38], that this amplitude should reach maximum at $K = L_D^{-1}$. This can also be used for estimation of the diffusion length $L_D \simeq 200$ nm ($\mu\tau \simeq 1.6 \times 10^{-8}$ cm^2/V).

As seen from Fig. 3.40 and Eq. (3.71) the resonant frequency depends on the value of applied dc field, and the increase of E_0 shifts the resonant maximum to lower frequencies. The amplitude of the resonance grows versus applied field nearly quadratically, which is another intrinsic feature of the non-steady-state photo-EMF excited in the regime with space-charge waves [38, 61].

The dependencies of the resonant frequency and signal amplitude on the light intensity were also observed experimentally. They demonstrated nearly linear dependencies in the range $I_0 = 0.88 - 3.5$ W/cm^2, which is well expected since $\omega_r \propto \tau_M^{-1} \propto \sigma_0 \propto I_0$ and $J^\omega \propto \sigma_0 \propto I_0$.

Reviewing the experimental dependencies $\omega_r(I_0, K, E_0)$ and $J^\omega(I_0, K, E_0)$ described in this section we can conclude, that the observed resonance results primarily from the excitation of space-charge waves, but not some other modes, e.g., drift waves [38, 61].

We have started our investigation of the non-steady-state photo-EMF in β-Ga_2O_3 using the light with the wavelength $\lambda = 532$ nm and photon energy $h\nu =$ 2.33 eV, while the crystal's band gap is of 4.8 eV. The origin of the photoeffect at chosen wavelength is not evident to date. The dark conductivity of Ga_2O_3 is usually attributed to the presence of oxygen vacancies, which are shallow donors [197]. The photoexcitation of these donors seems the most plausible explanation of the photoconductivity.

The necessity of crystal characterization in different spectral ranges is not the only reason of our choice of the green light instead of the commonly used UV. It seems sensible to begin investigation of a new material for the simplest case. Analyzing the experimental results we can confine ourselves to the case of monopolar photoconductivity for cw laser radiation with $\lambda = 532$ nm, but we would have to consider the bipolar one and estimate much more parameters, if the photon energy was sufficient for the interband transitions. Besides, the high energy photons would absorb in the thin surface layer making the problem to be two-dimensional with the rather bulky formulae and more complex fitting procedure [87]. Nevertheless, the excitation by the UV light is very promising, since the larger signal amplitude is expected.

It was noted above that the amplitude of the non-steady-state photo-EMF in Ga_2O_3 is inferior to that observed in sillenite crystals ($E_g = 3.15 - 3.47$ eV [2]) for the same experimental conditions. It is resulted essentially from the low photoconductivity in the visible light range. There are several parameters, that determine the value of the photoconductivity and the resulting signal amplitude, namely, the mobility and lifetime of charge carriers, absorption coefficient and quantum efficiency. The rather high value of the mobility $\mu = 46 - 83$

$cm^2V^{-1}s^{-1}$ [193, 194] partially compensates the low lifetime in the Ga_2O_3 crystal. The estimation of the absorption coefficient of $1.1 - 1.3$ cm^{-1} is comparable with the values reported earlier for Ga_2O_3 [195] and nearly coincide with the typical values for $Bi_{12}SiO_{20}$ at $\lambda = 532$ nm [2].

The studied Ga_2O_3 crystal has some advantages over the mentioned $Bi_{12}SiO_{20}$ crystal. The frequency transfer function of the former is flat in the range of $0.1 - 100$ kHz, while the signal in the latter decays over 10 kHz. Moreover, the additional measurements of the photoconductivity relaxation in the Ga_2O_3 crystal allow us to state that the frequency dependence should be flat up to 600 kHz. This means that the photoconductivity relaxation time, which is equal to the electron lifetime in the simplest case, should not exceed 0.26 μs. No influence of shallow traps has been revealed (compare Fig. 3.36 with those in Refs. [173, 174]). The material has a moderate static dielectric constant $\epsilon \simeq 10$, which is about 5 times lower than that in sillenite crystals, and this allows to achieve the appropriate cut-off frequency at low illumination levels.

Our investigations reveal the small difference of electrical properties along the [100] and [010] directions. The stronger anisotropy can be present for the other pairs of axes. For example, the difference of the conductivity and mobility along the [010] and [001] axes exceeding an order of magnitude was reported earlier in [193]. The larger conductivity and mobility along the [010] axis was attributed there to the crystal structure with the rutile chains arranged along [010], which results in substantial interaction among gallium $4s$ orbitals in the octahedral sites. However, the experiments in [194] suggest that the electrical conductivity almost does not depend on the crystallographic direction. In any case, the observation of the non-steady-state photo-EMF for all crystal orientations is considered as a task for the further studies.

The estimates of the material parameters obtained in the space-charge wave excitation regime should be considered as very rough. The resonant maxima have the shape different from the theoretically predicted, which can result from the nonuniformity of the dc electric field distribution. This nonuniformity arises in its turn from the voltage drop near the blocking electrode and from the inhomogeneous light distribution.

As it was mentioned above, gallium oxide grown with the oxygen deficiency demonstrates switching behavior, i.e., electrical nonlinearity [183]. The observation of the nonlinear optical phenomena such as second harmonic generation and photorefractive effect is prohibited in Ga_2O_3, since the crystal is centrosymmetric, and the third range tensors responsible for these effects vanish. Nevertheless, the nonlinear optical effects of higher orders are possible of course, e.g., the third harmonic generation and photorefractive effect via Kerr nonlinearity. Generally speaking, the studied non-steady-state photo-EMF effect is also a nonlinear phenomenon [174]. It appears as a product of joint action of the photoconductivity and space charge field effects, both of which arise under illumination.

In conclusion, we have demonstrated the capabilities of the non-steady-state photo-EMF technique in characterization of the wide-bandgap monoclinic Ga_2O_3 crystal. The signal was studied in the absence and in the presence of an external dc electric field. The photoelectric parameters such as photoconductivity and diffusion length of electrons are estimated. There is a weak anisotropy of the photo-EMF and electric parameters along the [100] and [010] axes and a pronounced polarization dependence of the quantum efficiency and resulting signal. The rather high amplitude of the signal and flatness of the frequency response in Ga_2O_3 makes this crystal a potential candidate for the fabrication of adaptive sensors of phase- and frequency-modulated optical signals.

Chapter 4

Non-Steady-State Photo-EMF and Two-Wave Mixing in Photorefractive Crystals under Frequency-Modulated Illumination

The evolution of photoinduced space charge under nonuniform and nonstationary illumination gives rise a number of nonlinear effects in photorefractive media: two- and four-wave mixing, space-charge waves, non-steady-state photo-EMF, etc. [2, 198]. Some of them have been observed with running or oscillating interference pattern. Such patterns are created by two coherent light beams shifted by frequency or phase modulated by a sinusoidal signal, respectively. Utilization of a light pattern running with constant velocity usually results in dc output signal, while the techniques using oscillating pattern provide ac one. Constant frequency shift can arise because of the Doppler effect and sinusoidal phase modulation appears in the beam reflected from a vibrating object. Corresponding schemes of velocimeters and vibrometers using photorefractive materials have been proposed [21, 139, 199–205].

In this study we try to extend the variety of the effects observed in photorefractive and photoconductive media introducing linear frequency modulation (LFM) in the light beams forming interference pattern. We investigate theoretically and experimentally the manifestation of such illumination in the non-steady-state photo-EMF and two-wave mixing effects. The conventional effect of the non-steady-state photo-EMF reveals itself as as alternating current arising in semiconductor illuminated by an oscillating interference pattern [19, 20, 38]. The electric current appears due to spatial shifts of the space-charge and photoconductivity distributions in the crystal volume. The related techniques realized with an interference pattern running with constant velocity is called as holographic current [131] or moving photocarrier grating technique [32]. The two-wave mixing effect is studied for years in photorefractive crystals and consists in intensity or phase redistribution for light waves propagating in the direction of the signal and reference beams [45, 63, 206–208]. In the case of oscillating interference pattern the amplitudes and phases of these waves become time-dependent because of the periodic spatial mismatch of the light intensity and refractive index distributions. In the case of running light pattern they observe dc two-wave mixing signal (stationary changes of light intensity behind the crystal). The excitation of both the photo-EMF and wave mixing signals involves processes of carrier generation, diffusion and drift in an electric field, recombination to local centers. This complex nature of the effects makes them powerful tools for determination of photoelectric parameters such as type and value of photoconductivity, carrier lifetime, mobility and diffusion length, density of recombination centers, etc. [32, 38, 132–134, 209].

One can expect that linear frequency modulation will result in pulsed response. Indeed, the response of the high-resistive material to initial fast variations of light intensity is negligibly small. At the moment when the interference pattern slows down and stops a noticeable space-charge grating is recorded, and signals can appear. Then the pattern begin to run fast again and signals vanish. In the following sections we consider this effect in details and discuss its possible applications.

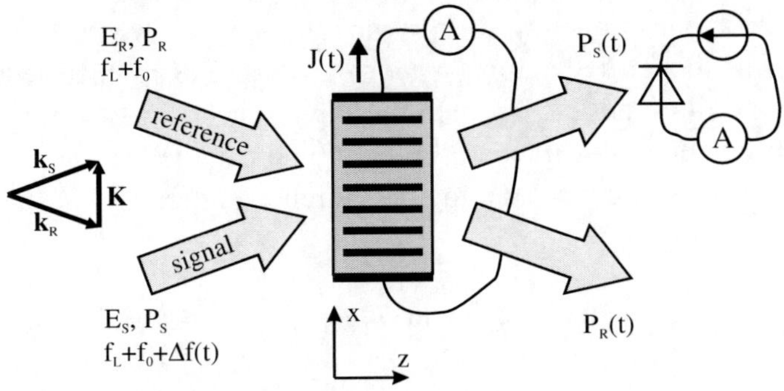

Figure 4.1. Scheme of the space charge formation and excitation of the non-steady-state photo-EMF and two-wave mixing signals by frequency modulated light.

1. Theoretical Analysis

The main idea of the experiment is sketched in Fig. 4.1. The crystal is illuminated by two light waves:

$$E_R(\mathbf{r},t) = (E_R/2)\exp[i\mathbf{k_R r} - i\psi_R(t)] + c.c., \tag{4.1}$$
$$E_S(\mathbf{r},t) = (E_S/2)\exp[i\mathbf{k_S r} - i\psi_S(t)] + c.c. \tag{4.2}$$

The both waves are polarized perpendicular to the incidence plane, the incidence angles are equal by modulus. These waves are frequency detuned by frequency shifters:

$$\psi_R = 2\pi(f_L + f_0)t, \tag{4.3}$$

$$\psi_S = 2\pi(f_L + f_0)t + 2\pi\int_0^t \Delta f(t')dt', \tag{4.4}$$

where $f_L = c/\lambda$ is the frequency of light, f_0 is the initial frequency shift ($\sim 80 - 120$ MHz for acoustooptic modulators), and $\Delta f(t)$ is the frequency shift in the signal beam, which results in signal appearance. We consider the case of linear frequency modulation:

$$\Delta f(t) = At. \tag{4.5}$$

We should note that frequency shift by an acoustooptic modulator can be accompanied with beam deflection: in our case this means that vector $\mathbf{k_S}$ can change not only its modulus but also the direction. However, we assume the frequency shifters to be ideal, operating without beam deflection.

These beams produce the running interference pattern:

$$I(x,t) = I_0\{1 + m\cos[Kx + \varphi(t)]\}, \tag{4.6}$$

where I_0 is the average intensity, m and K are the contrast and spatial frequency of the interference pattern,

$$\varphi(t) = -2\pi \int_0^t \Delta f(t')dt' = -\pi A t^2, \tag{4.7}$$

is the phase of the interference pattern. In fact, grating vector $\mathbf{K} = \mathbf{k_S} - \mathbf{k_R}$ and corresponding spatial frequency $K = (\mathbf{K})_x$ are time dependent since $\mathbf{k_S} = \mathbf{k_S}(t)$. We neglect these variations as they are very small: $|\Delta K|/K \simeq |\Delta f(t)|/2f_L \leq 10^{-8}$ for sensible frequency shifts $|\Delta f(t)| \leq 10$ MHz and not very small angles between beams $(0.1 - 1$ rad). Expression (4.6) can be rewritten in the form more convenient for calculations of space charge field:

$$I(x,t) = I_0\left[1 + \frac{m(t)}{2}e^{iKx} + \frac{m^*(t)}{2}e^{-iKx}\right], \tag{4.8}$$

where

$$m(t) = m\exp[i\varphi(t)]. \tag{4.9}$$

The further analysis is performed for the simplest model of electron semiconductor with one type of partially compensated donor centers [2]. We consider the diffusion mechanism of space charge formation, which takes place in the absence of an external electric field. The following assumption are applied as usually: the excitation and recombination of electrons are linear, dark conductivity and photovoltaic effect are absent, contrast m and light absorption are small. Then distributions of the density of electrons $n(x,t)$ and space charge field $E_{sc}(x,t)$ are sinusoidal and can be presented as follows [2]:

$$n(x,t) = n_0\left[1 + \frac{a(t)}{2}e^{iKx} + \frac{a^*(t)}{2}e^{-iKx}\right], \tag{4.10}$$

$$E_{sc}(x,t) = \frac{E_{sc}(t)}{2}e^{iKx} + \frac{E^*_{sc}(t)}{2}e^{-iKx}. \tag{4.11}$$

Complex amplitudes $a(t)$ and $E_{sc}(t)$ are found from the following set of linear differential equations which are just rewritten material, continuity and Poisson equations:

$$\frac{da}{dt} = -\frac{1+K^2L_D^2}{\tau}a + iK\mu E_{sc} + \frac{m(t)}{\tau}, \tag{4.12}$$

$$\frac{dE_{sc}}{dt} = -\frac{iE_D}{\tau_M}a - \frac{1}{\tau_M}E_{sc}. \tag{4.13}$$

Here μ, τ and L_D are the mobility, lifetime and diffusion length of electrons, $\tau_M = \epsilon\epsilon_0/\sigma_0$ is the Maxwell relaxation time of the material with the average conductivity $\sigma_0 = e\mu n_0$ and dielectric constant ϵ, $E_D = (k_BT/e)K$ is the diffusion field.

One can easily write down the solution of Eqs. (4.12), (4.13), but it is reasonable to make another simplifying assumption: formation of the space charge occurs with the quasi-stationary distribution of electrons in the conduction band [2], i.e. the electron lifetime τ is small compared to characteristic time of changes in the light pattern, in our case $\tau \ll |\Delta f(t)|^{-1}$. Then the desired solution is transformed to rather compact formulae:

$$a(t) = \frac{m(t) + i\mu\tau K E_{sc}(t)}{1+K^2L_D^2}, \tag{4.14}$$

$$E_{sc}(t) = \frac{-iE_D}{\tau_{sc}}\int_0^\infty m(t-t')e^{-t'/\tau_{sc}}dt'. \tag{4.15}$$

Now let us proceed with the calculation of the values measured in the experiment: photo-EMF and two wave mixing signals. The photo-EMF signal is defined by the drift component of the photocurrent averaged over interelectrode spacing and presented as combination of complex amplitudes $a(t)$ and $E_{sc}(t)$ [19,20,38]:

$$j(t) = \sigma_0 \,\mathrm{Re}[a(t)E_{sc}^*(t)]/2, \tag{4.16}$$

where Re means real part. After necessary substitutions we obtain the expression for photo-EMF signal:

$$j(t) = -\frac{m^2\sigma_0 E_D}{2(1+K^2L_D^2)}\,\mathrm{Im}\,\Xi(t), \tag{4.17}$$

where Im means imaginary part and

$$\Xi(t) = \tau_{sc}^{-1} \int_{0}^{\infty} \exp\{i[\varphi(t) - \varphi(t-t')] - t'/\tau_{sc}\} dt'. \tag{4.18}$$

The signal in two wave mixing experiments is usually detected by a photodiode, which current is proportional to the power of light emerging the crystal in the direction of signal (or reference) beam. If we neglect the absorption and reflection of the light and consider the case of thin holograms, then the power of "signal" beam behind the crystal is as follows [207, 208]:

$$\begin{aligned} P_S(t) &= P_S + P_R \delta^2 d^2 |E_{sc}(t)|^2/4 \\ &\quad + \sqrt{P_R P_S} \delta d \, \mathrm{Im}[e^{i\varphi(t)} E_{sc}^*(t)], \end{aligned} \tag{4.19}$$

where P_R and P_S are the powers of the incident light in reference and signal beams, $\delta = \pi n_0^3 r / \lambda' \cos\theta$ is the coefficient [45, 206], n_0 is the refractive index, r is the electrooptic coefficient, $\lambda' = c/(f_L + f_0)$ is the wavelength of the light after frequency shifter, 2θ is the angle between beams inside the crystal, d is the crystal thickness. The first term is the transmitted light of the signal beam, and it is time independent. The second term describes the light diffracted from the reference beam. It is weak because of the small diffraction efficiency of thin holograms: $|\delta d E_{sc}/2| \ll 1$. The last term determines the main part of the two wave mixing signal, and after necessary substitutions it equals

$$P_{S\sim}(t) = P_0 \frac{m^2 \delta d E_D}{2} \mathrm{Re}\, \Xi(t), \tag{4.20}$$

where $P_0 = P_R + P_S$ is the total power of the incident light. The two wave mixing signal in the "reference" beam behind the crystal has the opposite sign: $P_{R\sim}(t) = -P_{S\sim}(t)$.

Fortunately, for the considered linear frequency modulation (4.7) integral (4.18) can be represented via complementary error function:

$$\begin{aligned} \Xi(t) &= \frac{1}{2\sqrt{A\tau_{sc}^2}} \exp\left[-\frac{t}{\tau_{sc}} + i\left(\frac{\pi}{4} + \frac{1}{4\pi A \tau_{sc}^2} - \pi A t^2 \right) \right] \\ &\quad \times \mathrm{Erf}\left(-\sqrt{\frac{\pi A}{i}} t + \sqrt{\frac{i}{4\pi A \tau_{sc}^2}} \right). \end{aligned} \tag{4.21}$$

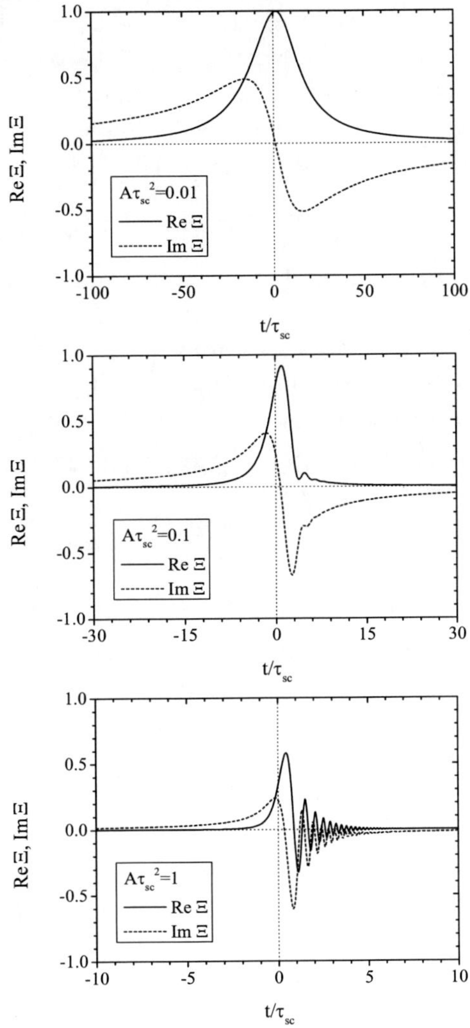

Figure 4.2. Time dependencies of the real and imaginary parts of integral Ξ (normalized two-wave mixing and photo-EMF signals) calculated for different sweep rates $A\tau_{sc}^2 = 0.01, 0.1, 1$.

This allows us to calculate time dependencies of the photo-EMF and two wave mixing signals (Fig. 4.2). Having rewritten Ξ for dimensionless time t/τ_{sc} one can easily note that this function has the only parameter $A\tau_{sc}^2$.

Solution (4.21) can be simplified for two extreme cases. The former corresponds to very slow frequency modulation ($\pi A\tau_{sc}^2 \ll 1$):

$$\Xi(t) = (1 + i2\pi At\tau_{sc})^{-1} = (1 + i2\pi\Delta f(t)\tau_{sc})^{-1}. \tag{4.22}$$

This result is exactly the same as that calculated for the case of constant frequency shift $\Delta f(t) = \Delta f_c = \mathrm{const}(t)$ [200]. This means, in particular, that frequency dependence measured by spectrum analyzer (with lowest sweep rate) coincides with the frequency dependence measured by "point by point".

Another extreme case realizes at fast frequency modulation ($\pi A\tau_{sc}^2 \gg 1$). For $t < -\tau_{sc}$ function $\Xi(t)$ reduces to Eq. (4.22) as well. For $t > \tau_{sc}$ an additional oscillatory behavior appears:

$$\begin{aligned} \Xi(t) &= (1 + i2\pi At\tau_{sc})^{-1} + (A\tau_{sc}^2)^{-1/2} \\ &\quad \times \exp[-t/\tau_{sc} + i(\pi/4 - \pi At^2)]. \end{aligned} \tag{4.23}$$

The appearance of oscillations can be simply explained. The interference pattern runs very fast at $t < 0$ preventing formation of a space charge field grating. Then it slows down and stops at $t = 0$. The stationary field grating with a noticeable amplitude arises. At $0 < t < \tau_{sc}$ the interference pattern begins to run in the opposite direction but the recorded stationary grating is still present. This combination of running interference pattern and stationary grating gives rise the oscillations of the current and light intensity analogous to the ones in experiments with conventional non-steady-state photo-EMF and two-wave mixing. These oscillations are decaying since the space charge field grating erases with characteristic time τ_{sc}.

Evaluating the real and imaginary parts of Eq. (4.22) one can note difference in their behavior at large t (Fig. 4.2): the real part (two-wave mixing signal) decays as t^{-2} while the imaginary one (photo-EMF signal) decays as t^{-1}.

2. Experimental Results

The experimental setup used for measurements of the non-steady-state photo-EMF and two-wave mixing signals is shown in Fig. 4.3. The light from He-Ne

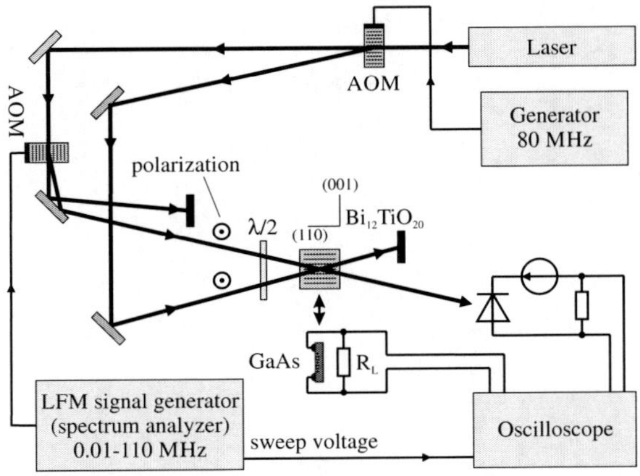

Figure 4.3. Experimental setup used for measurements of the non-steady-state photo-EMF and two-wave mixing signals excited by frequency modulated light. AOM are the acoustooptic modulators. The sample (GaAs or $Bi_{12}TiO_{20}$) is placed in the intersection of light beams in the corresponding experiment.

laser ($\lambda = 633$ nm, $P_{out} \simeq 30$ mW) passes through two acoustooptic modulators ML-201-1, which introduce frequency shifts f_0 and $f_0 + \Delta f(t)$ into diffracted light beams ($f_0 = 80$ MHz). We use the output of spectrum analyzer SK4-59 as a generator of the linear frequency modulated signal because of its wide range of frequency deviation. Then these beams are directed to the sample forming the interference pattern with average intensity $I_0 = 240$ mW/cm^2 and contrast $m = 0.97$ (the power of signal and reference beams is $P_S = 3.4$ and $P_R = 2.0$ mW, respectively). The excitation of the non-steady-state photo-EMF is carried out in semiinsulating GaAs crystal. The sample has dimensions of $3 \times 3 \times 0.5$ mm^3, the front and back surfaces 3×3 mm^2 were polished. Two gold stripes were deposited on the front surface (with interelectrode spacing of 1 mm). The sample is tilted around the direction of light propagation by $\sim 15°$ in order to decrease contact photo-EMF [210]. The non-steady-state photo-EMF creates voltage on the load resistor $R_L = 1$ MΩ which is measured by the digital oscilloscope. The experiments on two-wave mixing are performed with $Bi_{12}TiO_{20}$ undoped crystal ($9.5 \times 12 \times 11$ mm^3). The sample has one of the standard holographic orientations: cut (110) with axis [001] lying in the in-

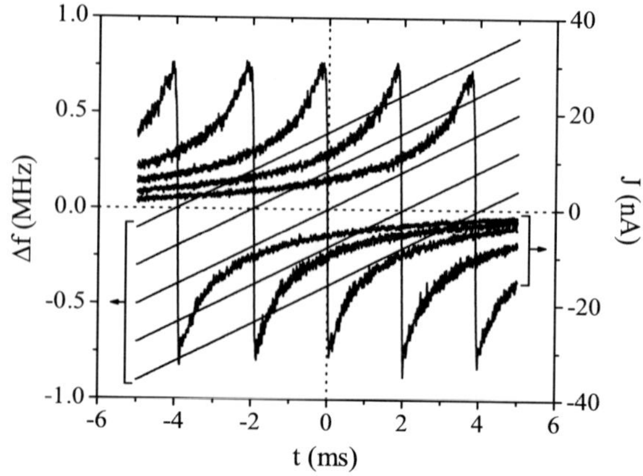

Figure 4.4. Oscillograms of the frequency shift in signal beam and the corresponding current arising in GaAs crystal. $\lambda = 633$ nm, $I_0 = 240$ mW/cm^2, $K = 190$ mm^{-1}, $A = 10^8$ Hz/s.

cidence plane. The front and back surfaces 9.5×12 mm^2 were polished, no electrodes were deposited. Since the crystal is rather thick and rotates polarization by $\sim 70°$, the halfwave plate is placed in front of it. As a result, the polarization plane in the middle of the crystal is settled perpendicular to the incident plane. The power of the light emerging the crystal in the direction of the signal beam is measured by the photodiode and visualized by the oscilloscope. The sweep voltage from spectrum analyzer feeds the oscilloscope as well allowing simultaneous measurements of the frequency shift and investigated signal.

First let us find out if the moment of signal appearance is unambiguously determined by the slowing-down and stopping of the interference pattern, i.e. by the coincidence of frequencies of the signal and reference beams. We settle several different central frequencies on spectrum analyzer ($79.6 - 80.4$ MHz) and measure corresponding traces of the non-steady-state photo-EMF (Fig. 4.4). As seen the pulse position, i.e. the time when photocurrent changes its sign, well corresponds to the moments when frequencies of light waves are equal ($\Delta f = 0$).

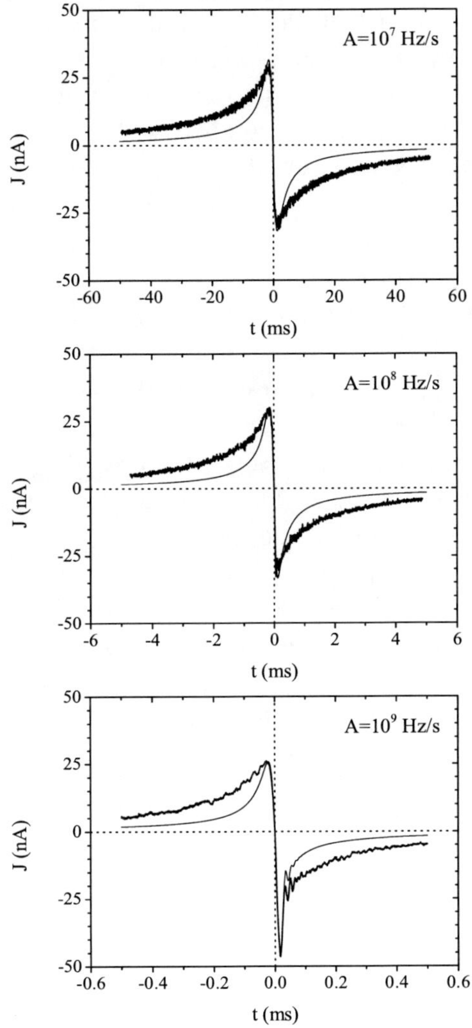

Figure 4.5. Oscillograms of the photocurrent in GaAs crystal measured for different sweep rates $A = 10^6, 10^7, 10^8, 10^9$ Hz/s. $\lambda = 633$ nm, $I_0 = 240$ mW/cm^2, $K = 190$ mm^{-1}. Thin blue lines show approximation by Eq. (4.21) for $\tau_{sc} = 12\ \mu$s.

Figure 4.5 presents traces of the non-steady-state photo-EMF signal measured at different sweep rates. The signal demonstrates qualitatively the same behavior as described in theoretical analysis. At low sweep rates the pulse consists of two parts with nearly equal amplitudes and opposite signs. The increase of sweep rate decreases the pulse duration, while the amplitude remains nearly constant. At large sweep rates the negative peak becomes sharper than positive one. At very large A chirped and damped oscillations appear. The most pronounced oscillations are achieved by simultaneous increase of sweep rate and decrease of illumination level (the pulse amplitude and corresponding signal-to-noise ratio become lower, however).

There are, however, some peculiarities which does not described by the developed theory. First, the pulse measured at lowest sweep rate ($A = 10^6$ Hz/s) should be the most symmetric. In practice the preceding positive half has 1.9 times smaller amplitude than the following negative one (trace is not shown in the figure). Second, the pulse tails decay slower than t^{-1} as predicted by Eq. (4.22). The thin blue lines in Fig. 4.5 show the approximation by Eq. (4.21) for $\tau_{sc} = 12\ \mu$s. The origin of these discrepancies is unknown, we just can suppose that it can be associated with bipolar conductivity and high light absorption in GaAs at $\lambda = 633$ nm. The both factors sufficiently change the dynamics of space charge formation and non-steady-state photo-EMF characteristics [86, 87, 211].

We have also measured the dependence of the signal on the light intensity (Fig. 4.6). The peak-to-peak amplitude of the pulse demonstrate noticeable sublinearity $J_{p-p} \propto I_0^{0.56}$ in range $I_0 = 11 - 240$ mW/cm^2, what is rather typical for GaAs crystal at this wavelength. Decreasing the light intensity we observed similar evolution of the pulse shape as that for increasing sweep rate. This fact confirms the statement that pulse shape is determined by parameter $A\tau_{sc}^2$.

Traces of the two-wave mixing signals excited by LFM light waves are shown in Fig. 4.7. Chirped pulses are observed for sweep rates of $0.25 - 50$ kHz/s. Their behavior is well described by the developed theory: the pulses appears at the moments when the interference pattern stops, the oscillations are frequency modulated and damped, their duration and amplitude decreases with increasing sweep rate. Nevertheless the signals are noisy, we can state that the pulse tails of two-wave mixing decay faster than the ones of photo-EMF (at least

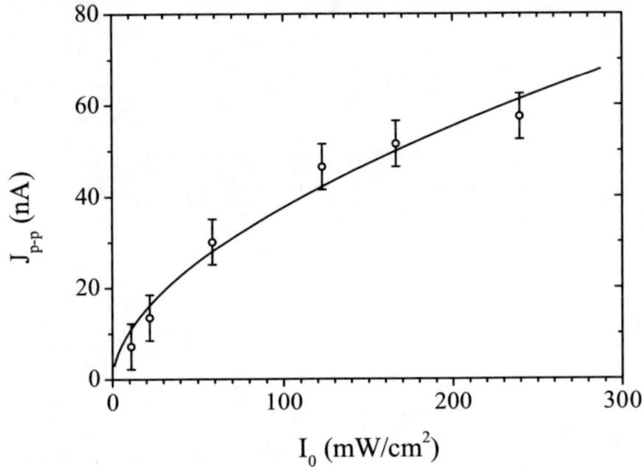

Figure 4.6. Dependence of the photo-EMF peak-to-peak amplitude versus average light intensity. GaAs, $\lambda = 633$ nm, $K = 190$ mm^{-1}, $A = 5 \times 10^7$ Hz/s. Solid line shows approximation $J_{p-p} \propto I_0^{0.56}$.

for $t < 0$), as the theory predicts.

The measurements are carried out at sweep rates much lower than those for photo-EMF experiments. Small photoconductivity of $Bi_{12}TiO_{20}$ crystals results in large time of space charge formation $\tau_{sc} \sim 0.1$ s. This means that even the lowest used sweep rate of 250 Hz/s satisfies condition $\pi A \tau_{sc}^2 \gg 1$ and that is why only oscillatory signals are detected. We can settle lower sweep rates but the signal becomes very unstable: the used vibroisolation is not enough for pulse durations larger than 100 ms.

We should note here that the theoretical analysis has been carried out for thin hologram, while the two-wave mixing experiment is realized in geometry of thick one ($\lambda d / n_0 \Lambda^2 \simeq 400 \gg 1$, where $\Lambda = 2\pi/K$). We have to use thick crystal because the signals are rather small and detected in wide band (5 kHz). We can estimate the maximal amplitude of two-wave mixing signal from Eq. (4.20): $P_0 m^2 \delta d E_D / 2 = 470$ μW. As seen this estimation exceeds the experimental value nearly by an order of magnitude. Several factors can be taken into account to explain this discrepancy: first, we measure only oscillating signals while the maximum amplitude should be achieved at lower sweep rates, second, the approach of thin hologram provides overrated diffraction efficiency

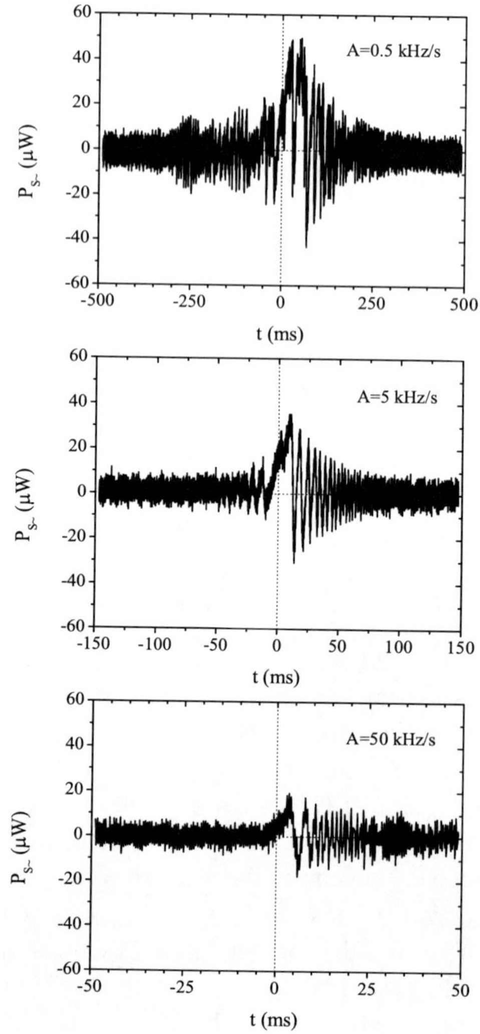

Figure 4.7. Oscillograms of the two-wave mixing ac signal measured in $Bi_{12}TiO_{20}$ crystal at different sweep rates: $A = 0.5, 5, 50$ kHz/s. $\lambda = 633$ nm, $I_0 = 240$ mW/cm^2, $K = 2.3\ \mu m^{-1}$.

for thick one [2], third, there are noticeable light absorption and optical activity which decrease the signal as well.

The results of theoretical analysis and experiments in GaAs and $Bi_{12}TiO_{20}$ generally match each other. The time of signal appearance, variation of the pulse duration and shape versus sweep rate and light intensity are well described by the theory. However, some features such as pulse tails in GaAs and low pulse amplitude in $Bi_{12}TiO_{20}$ require more advanced theoretical approaches dealing with bipolar conductivity, high light absorption, vectorial coupling [86,87,211–214].

For years we emphasized the fact that frequency transfer function of the conventional non-steady-state photo-EMF measured at low frequencies is similar to that of simplest differentiating RC-circuit (high-pass filter) [20, 38]. In the case of LFM signals this similarity vanishes: at large $|t|$ the frequency is large too, and ac voltage freely passes through differentiating RC-circuit; at low $|t|$ when the frequency is small the signal is damped. We can note that behavior of the photo-EMF signal for linear frequency modulation is more similar to the one of the integrating RC-circuit (low-pass filter). Indeed, the voltage $\mathrm{Re}\{U_0 \exp[i\varphi(t)]\}$ feeding low-pass RC-filter produces the following output signal:

$$U(t) = \mathrm{Re}\left\{ \frac{U_0}{RC} \int_0^\infty \exp[i\varphi(t - t') - t'/RC]dt' \right\}, \tag{4.24}$$

what reminds the signals investigated in this paper [Eq. (4.18)].

Let us point out the possible applications of the studied effects. The pulses appear at the moments when frequencies of the signal and reference beams coincide. This fact can be used in Doppler scanning velocimeter estimating the unknown velocity from the time delay of the detected pulses. Moreover, the duration and shape of the pulses provide information about sweep rate of the LFM signal, which can be caused by the uniformly accelerated motion of the object. This means that the velocity and acceleration can be measured simultaneously by detection of the only pulse, what is very actual in investigations of fast processes. In this work we used frequency shifts and sweep rates in ranges $|\Delta f| = 0 - 10$ MHz and $|A| = 0.0005 - 1000$ MHz/s. They would correspond to the following ranges of velocity and acceleration of the light reflecting object: $|V| = 0 - 3$ m/s and $|a| = 0.0003 - 600$ m/s^2.

The devices realized on the considered effects will inherit the advantages of the devices using conventional non-steady-state photo-EMF and two-wave mixing effects. They include the rather high sensitivity and adaptivity, i.e. the possibility of operation in the presence of a slow phase drift in the interferometric setup. The devices will also be able to detect light signals with complicated wave fronts, even with speckle patterns produced by reflection from the real diffusely scattering objects. However, these new methods will have some limitations typical for the standard techniques [215, 216]: the light source should possess enough transverse and longitudinal coherence to form an interference pattern with maximum contrast; large spatial frequency K may also be preferable, so that each speckle spot contains several interference fringes [135].

The developed techniques can also be used for characterization of widegap semiconductors. Instead of investigations of frequency transfer functions we can measure the pulse response and estimate space-charge formation time τ_{sc} and corresponding photoconductivity σ_0. Both the low and high sweep rates of the LFM signal can used for estimation of τ_{sc} [see Eqs. (4.22) and(4.23)].

Chapter 5

Adaptive Interferometers using the Effect of the Non-Steady-State Photoelectromotive Force and Two-Wave Mixing in Photorefractive Crystals

Detection of low-amplitude acoustic vibrations of real objects, such as ultrasonic transducers, composite materials, biological objects, metal production, MEMS is an important scientific and technical problem [6–13]. Homodyne laser vibrometers [7] are suitable for practical applications in this regard for their high sensitivity which is limited in principle only by shot-noise of the laser used. These vibrometers allow detection at a distance, with high spatial localization of the measurement region and wide frequency range. However, their utilization is restrained by several problems such as slow phase drifts in the interferometer arms due to environmental reasons, necessity of fine optical adjustment and suppression of laser amplitude noise.

Photorefractive materials (and in particular semiconductor GaAs, InP:Fe and CdTe) offer a novel elegant way to solve the problem of keeping operation interferometer point constant [2, 14–17]. The crystal replaces the conventional beamsplitter and can be controlled not only electrically but also optically, i.e., based on the principles of nonlinear optics. In addition, it can be multilayered. Such a multilayered adaptive beamsplitter is nothing else but a volume dynamic hologram recorded by the reference and signal waves.

Adaptive detectors based on the effect of the non-steady-state photo-electromotive force (photo-EMF) are promising for sensor applications [18]. This optical phase-to-electrical signal converter is not sensitive to ambient vibrations, thermal drift, amplitude laser noise and allow detection of speckle-like patterns.

The nonstationary holographic photocurrent [19] (or non-steady-state photo-EMF [20]) is a holography related effect and it appears in a semiconductor material illuminated by an oscillating light pattern. Such illumination is usually created by two coherent light beams one of which is phase modulated with frequency ω. The alternating current is resulted from the periodic relative shifts of the photoconductivity and space charge gratings which arise in the crystal's volume under illumination. Like the holographic recording in photorefractive crystals this effect demonstrates adaptive properties that promote its application in such areas as vibration monitoring, velocimetry, etc. [21]. The sensitivity of adaptive photodetector is comparable with the sensitivity of classical schemes using photodiodes. The amplitude of the non-steady-state photo-EMF can be increased by an external ac field so one can expect the improvement of the sensitivity in certain ranges of temporal and spatial frequencies. In contrast to the holographic methods this technique allows the direct transformation of phase modulated optical signals into the electrical current and can be applied for characterization of centrosymmetrical and even amorphous materials. Since the photocurrent is originated from the interaction of both the photoconductivity and space charge gratings a lot of photoelectric parameters can be measured [20].

1. Non-Steady-State Photo-EMF in an Interferometer with Optoelectronic Feedback

Among the number of the nonlinear phenomena observed in photorefractive crystals there are effects which appearance or specific behavior is determined by additional feedback loop. Stabilization of a holographic setup was one of the earliest and the most important application of the optoelectronic feedback [207]. It is implemented using the detection scheme and the piezoelectric mirror placed in one of the arms of the setup. The mirror is driven by the amplified voltage of the detector providing the necessary feedback loop. Such feedback keeps the difference of the optical path lengths to be constant and correspondingly stabilizes position of the interference fringes. As a result the holograms are recorded with high modulation index.

Further investigations of the feedback operation have revealed dramatic changes in photorefractive recording itself. It was found that the self-stabilized recording in $LiNbO_3$ leads to an almost 100% diffraction efficiency of the hologram [217, 218]. It was shown later that this regime is quasi-steady state: the diffraction efficiency oscillates in the close vicinity of 1 (or 0) and the phase of the controlled beam grows almost linearly [219, 220]. These effects are accompanied by the light scattering reduction [221] and fringe's bending [219]. Besides, the feedback controlled two-wave coupling can result in the excitation of space charge waves [222]. A number of nonlinear oscillation regimes is predicted in Ref. [223] for materials with local and nonlocal photorefractive responses and different phase relations between diffracted and transmitted beams.

Generally speaking, the implementation of the feedback in optical schemes reveals features typical to those in common dynamical systems (mechanical, electronic, biological) considered in the theory of oscillation and control theory [224, 225]. For example, the negative feedback in Michelson interferometer provides the ability to linearize the sensor (photodiode) output allowing detection of large displacements [226]. The conventional electronic operational amplifier demonstrates similar behavior with the feedback loop. Various optical schemes with positive feedback such as laser, molecular cell inside Fabry-Perot resonator, hybrid schemes with birefringent materials [227] reveal the oscillatory or bistable behavior which also has analogs in electronics and mechanics (harmonic oscillators, triggers). Since the presence of the properly chosen feedback in dynamical system not only improves its performance but can also provide observation of new nonlinear phenomena, we have applied feedback operation concept to the sensor based on the effect of the non-steady-state photo-EMF. In contrast to techniques developed in Refs. [207, 217–222] we do not use any dither signals and synchronous detection. Our approach implies the utilization of a conventional wideband amplifier in the electronic circuit. Both the diffusion and drift regimes are investigated theoretically and experimentally in $Bi_{12}SiO_{20}$ crystal.

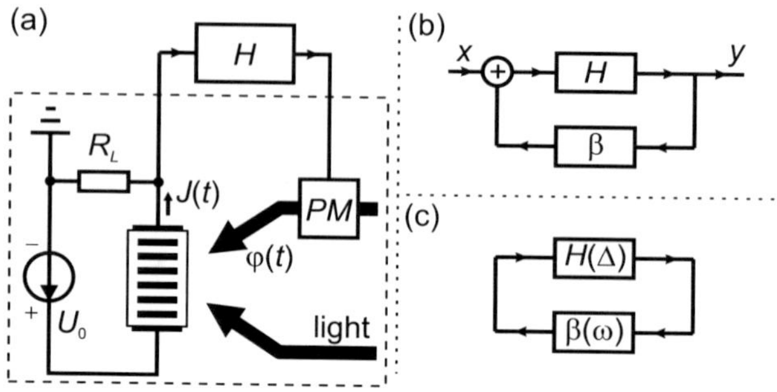

Figure 5.1. Block diagrams illustrating closed-loop operation of the interferometric scheme studied in this chapter (a), general feedback structure (b), and general oscillator scheme (c). PM is the phase modulator.

Let us start the analysis of our interferometric scheme (Figure 5.1, a) with formulation of the problem in terms of oscillation theory and control theory [224–226]. Dynamical systems are often represented by block diagrams, which have a unique algebra and set of transformations. The block diagrams are used because they are a shorthand pictorial representation of the cause-and-effect relationship between the input and output in a real system, and they are a convenient method for characterizing the functional relationships between components. One important arrangement is the general feedback form illustrated in Figure 5.1, b. In this figure H and β represent the transfer functions from the input to the output of the systems contained within their respective boxes. The overall transfer function is as follows:

$$T = \frac{H}{1 - H\beta}. \tag{5.1}$$

Product $H\beta$ is often called loop gain. When this product tends to unity the transfer function diverges what corresponds to the appearance of an instability and unlimited growth of oscillations in the closed-loop system.

The basic form of a harmonic oscillator is an amplifier with the output attached to a narrow-band filter [$\beta = \beta(\omega)$], and the output of the filter attached back to the input of the amplifier (Figure 5.1, c). The unlimited growth of oscillations is usually restricted by nonlinearity of the amplifier observed at large amplitudes, i.e., $H = H(\Delta)$. Relation

$$H(\Delta)\beta(\omega) = 1 \tag{5.2}$$

is the main equation (Barkhausen equation) defining the amplitude Δ_g and frequency ω_g in the steady-state oscillation regime. So, our main goal in this section is to obtain the analogous equation for the interferometric scheme using the non-steady-state photo-EMF detector and feedback loop (Figure 5.1, a) and calculate the corresponding oscillation amplitudes and frequencies.

Let us assume the crystal is illuminated by an oscillating interference pattern created by two light beams one of which is phase modulated with amplitude Δ and frequency ω:

$$I(x,t) = I_0[1 + m\cos(Kx + \Delta\cos\omega t)]. \tag{5.3}$$

Here I_0 is the average intensity, m and K are the contrast and spatial frequency of the interference pattern, Δ and ω are the amplitude and frequency of phase modulation.

Such illumination excites an ac current in the crystal's volume via non-steady-state photo-EMF effect. This current produces the corresponding voltage at load resistor, which is then amplified and transferred into phase modulation with amplitude Δ using the electrooptic modulator:

$$\Delta = \frac{JR_L H}{U_\pi/\pi}, \tag{5.4}$$

where U_π is the half-wave voltage of the phase modulator, R_L is the load resistance, and H is the gain factor of the amplifier. So the feedback loop is closed.

We can easily rewrite Eq. (5.4) in the form analogous to Barkhausen equation (5.2):

$$H\left(\pi\frac{J^\omega(\omega,\Delta)R_L}{U_\pi\Delta}\right) = 1, \tag{5.5}$$

where J^ω is the complex amplitude of the photo-EMF signal in the small-signal approximation. As seen, the expression in the parentheses plays role of feedback transfer function β in the standard Barkhausen equation (5.2).

We shall consider the ideal amplifier with the flat frequency transfer function and infinite dynamic range:

$$H = \text{const}(\omega, U_{in}). \tag{5.6}$$

In this case the frequency and amplitude of oscillations is defined by the frequency response and nonlinearity of the effect of the non-steady-state photo-EMF but not by the amplifier.

They usually differ two basic mechanisms of the space charge formation – diffusion and drift ones [2]. Let us proceed the analysis of the former.

For diffusion mechanism of recording no external voltage is applied to the sample and the amplitude of the non-steady-state photocurrent excited in the crystal with electron conductivity is given by the following expression: [20, 38]

$$J^{\omega} = \frac{-Sm^2 \mathrm{J}_0(\Delta)\mathrm{J}_1(\Delta)\sigma_0 E_D i\omega\tau_M}{1 - \omega^2\tau\tau_M + i\omega[\tau + \tau_M(1 + K^2 L_D^2)]}, \tag{5.7}$$

here S is the electrode area, σ_0 is the average photoconductivity of the sample, E_D is the diffusion field, τ_M is the Maxwell relaxation time, τ is the relaxation time of the conductivity, L_D is the diffusion length of electrons, and $\mathrm{J}_n(\Delta)$ is the Bessel function of the first kind of the n-th order.

First let us consider the case of real gain factor:

$$H = H_r \tag{5.8}$$

and

$$\text{Im}(H_r) = 0. \tag{5.9}$$

This situation is typical for conventional analog wideband amplifiers.

The phase conditions, i.e., phase shifts of the amplifier and feedback, define the operating frequency of generators in the oscillation theory. Namely, the sum of amplifier and feedback phase shifts must be multiple of 2π. We can always settle the phase of Δ to zero by the appropriate choice of the time origin. As follows from Eq. (5.7) the complex amplitude of the photocurrent is real value for modulation frequency equal to

$$\omega_g = \omega_m = (\tau\tau_M)^{-1/2}. \tag{5.10}$$

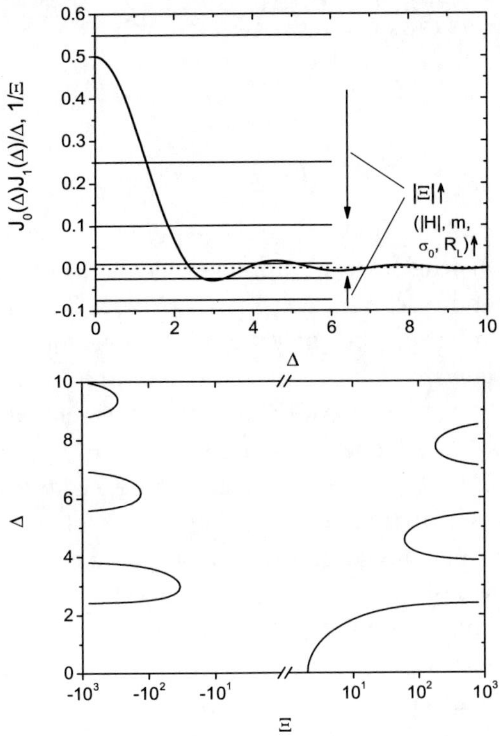

Figure 5.2. Graphic solution of Eq. (5.11) and corresponding dependence of the oscillation amplitude on the normalized gain factor.

It corresponds to the maximum in the frequency dependence of the photo-EMF amplitude.

The amplitude of oscillations in the conventional generators (electrical, optical, etc.) is usually defined by the amplifier's nonlinearity originating for large amplitudes. In our case the saturation of the non-steady-state photo-EMF for large Δ will confine the generation amplitude. As follows from Eqs. (5.5) and (5.7) it is defined as:

$$\mathrm{J}_0(\Delta)\mathrm{J}_1(\Delta)/\Delta = 1/\Xi. \tag{5.11}$$

Here we introduced the dimensionless gain factor Ξ:

$$\Xi = -\xi H_r E_D, \tag{5.12}$$

$$\xi = \frac{\pi S m^2 \sigma_0 R_L}{U_\pi (1 + K^2 L_D^2)}. \tag{5.13}$$

Equation (5.11) is nonlinear and can be solved graphically or numerically (Figure 5.2). Oscillation with lowest amplitude occurs when dimensionless gain factor Ξ is positive and reaches 2, i.e., $\Xi > 2$ (Figure 5.2). Increasing the gain factor from this value one can magnify the oscillation amplitude up to $\Delta \simeq 2.4$. When gain factor reaches $\Xi \simeq 60$, 180, etc. oscillations with higher amplitudes ($\Delta \simeq 4.6$, 7.8, etc.) become possible as well. As seen, each of these higher modes splits into two sub-modes with higher and lower amplitudes. Another feature is also seen from Figure 5.2. The first mode (with lowest amplitude) is characterized by the soft regime of generation, i.e., such generation can appear as the result of very small deviation from the unstable equilibrium $\Delta = 0$ (e.g., from the noise of load resistor). Higher modes are characterized by hard regime of generation and require sufficient initial deviation, which should be of order of the oscillation amplitude, i.e., ~ 4.6, 7.8, etc.

The generation is also possible for negative gain factors $\Xi < 0$ (Figure 5.2). In this case oscillations with lowest amplitude $\Delta \simeq 3.0$ appear when gain factor reaches $\Xi \simeq -34$. The behavior of modes is analogous to the one for positive gain factors. All the modes are characterized by hard regime of generation.

Now let us suppose that amplifier of the photo-EMF signal both increases the amplitude and introduces $\pi/4$ phase shift:

$$H = H_r \exp(i\pi/4). \tag{5.14}$$

Then the generation frequency equals

$$\omega_g = \omega_1 = \tau_{sc}^{-1} = [\tau_M (1 + K^2 L_D^2)]^{-1}. \tag{5.15}$$

It should be noticed that this estimation is obtained for condition $\tau \ll \tau_M$ which is typical for high resistive semiconductors (e.g., for photorefractive $Bi_{12}SiO_{20}$). Frequency ω_1 is the first cut-off frequency of the non-steady-state photo-EMF for the case of forced excitation of the signal [20] and it is the inverse value of the space charge formation time τ_{sc} [2].

The amplitude of oscillation is calculated from Eq. (5.11) with

$$\Xi = -\xi H_r E_D/\sqrt{2}. \tag{5.16}$$

Coefficient $\sqrt{2}$ appears here as the photo-EMF amplitude at frequency ω_1 is $\sqrt{2}$ times lower than that at frequency ω_m. Since the amplitude of oscillations Δ is defined from the same equation as for the case of real H then its behavior with respect to gain factor Ξ should be analogous to the one described above.

At last let the gain factor equal

$$H = H_r \exp(-i\pi/4). \tag{5.17}$$

Then the oscillation frequency corresponds to the second cut-off frequency of the photo-EMF signal:

$$\omega = \omega_2 = (1 + K^2 L_D^2)/\tau \tag{5.18}$$

and amplitude is deduced from Eqs. (5.11) and (5.16).

For drift mechanism of recording when an external dc electric field E_0 is applied to the crystal and the photo-EMF amplitude is defined by the following expression: [38]

$$\begin{aligned} J^\omega &= \frac{0.5Sm^2 \mathrm{J}_0(\Delta)\mathrm{J}_1(\Delta)\sigma_0[i2E_0 - \omega\tau_M(E_0 + iE_D)]}{1 - \omega^2\tau\tau_M + i\omega[\tau + \tau_M(1 + K^2L_D^2 + iKL_0)]} \\ &- \frac{0.5Sm^2 \mathrm{J}_0(\Delta)\mathrm{J}_1(\Delta)\sigma_0[i2E_0 - \omega\tau_M(E_0 - iE_D)]}{1 - \omega^2\tau\tau_M + i\omega[\tau + \tau_M(1 + K^2L_D^2 - iKL_0)]}, \end{aligned} \tag{5.19}$$

First let us suppose the gain factor is real [Eqs. (5.8), (5.9)]. Besides, let us consider the case of rather large electric fields such that conditions $E_0 \gg E_D$, $KL_0 \gg 1 + K^2L_D^2$ are satisfied. Then the generation frequency is equal to the first resonant frequency of the non-steady-state photo-EMF excited in forced regime [38] and it is equal to the frequency of space charge wave (trap recharging wave) – one of the fundamental modes of the space charge oscillation in semiconductors: [2]

$$\omega_g = \omega_{r1} = \omega_{scw} = (\tau_M K L_0)^{-1}. \tag{5.20}$$

The amplitude of oscillations is deduced from Eq. (5.11) with

$$\Xi = \xi H_r E_0 K L_0. \tag{5.21}$$

As follows from Eqs. (5.12) and (5.21) the generation mode is achieved easier in the drift regime, namely, the required gain factor of the amplifier H is $E_0 K L_0 / E_D$ times smaller than that for the diffusion regime. It is due to the effect of the resonant amplification of the running space charge gratings in an external dc field. Moreover the sign of the non-steady-state photo-EMF signal in the diffusion and drift regimes differs by π so the necessary phase shifts of the amplifier are opposite for these cases.

Now let us assume the gain factor is purely imaginary:

$$H = H_r \exp(i\pi/2). \tag{5.22}$$

Then generation frequency equals

$$\omega_g = \omega_{r2} = \omega_{dw} = K\mu E_0. \tag{5.23}$$

Frequency ω_{r2} is the second resonant frequency of the non-steady-state photo-EMF effect [38] and it is equal to the frequency of another eigenmode of space charge oscillations – drift wave [228]. The amplitude of oscillations is deduced from Eq. (5.11) with

$$\Xi = \xi H_r E_0/2. \tag{5.24}$$

We considered theoretically the self-excitation regime for the non-steady-state photo-EMF effect. The results of the analysis shows the possibility of materials characterization using this technique. The generation frequency is defined by the characteristic times of the space charge and conductivity relaxation. In contrast to the forced regime of the signal excitation the considered technique does not require measurements of the frequency transfer function, the oscillation frequency can be measured by conventional frequency counter.

The scheme of the experimental arrangement is presented in Figure 5.3. The second harmonic of Nd:YAG laser was used as the source of the coherent radiation ($\lambda = 532$ nm, $P_{out} \simeq 20$ mW). The light was split into two beams. One of them was phase modulated by electrooptic modulator (the light passed two times through the modulator to increase the efficiency of modulation, $U_\pi =$ 193 V). Then these two beams were directed onto crystal's surface where they

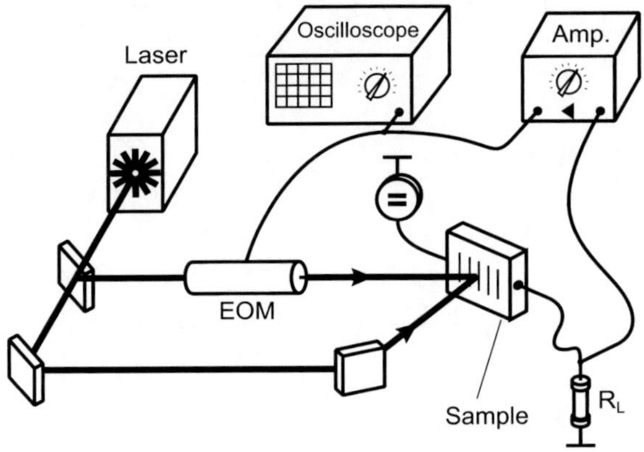

Figure 5.3. Simplified scheme of the experimental setup. EOM is the electrooptic modulator, Amp is the wideband amplifier with regulated gain factor.

produce oscillating interference pattern with the following parameters: $I_0 = 0.75$ W/cm^2, $m = 0.95$, $K = 1.0\ \mu$m^{-1} for the diffusion regime of signal excitation and $I_0 = 0.10$ W/cm^2, $m = 0.17$, $K = 0.16\ \mu$m^{-1} for the drift one. The current arising in the crystal under such illumination produce voltage on the load resistor ($R_L = 110$ kΩ and 10 kΩ for diffusion and drift mechanism of signal excitation, respectively). Then this voltage was amplified, measured by the analog-to-digital converter connected to computer, and applied back to the electrooptic modulator. The amplifier is characterized by the gain factor $40 - 80$ dB, phases 0 and π, flat frequency response in range 1.5 Hz – 150 kHz, and highest output voltage of 100 V. In our experiments we used conventional undoped $Bi_{12}SiO_{20}$ crystal. The sample has the following dimensions: $10\times4\times1$ mm^3. The silver paste electrodes (3×3 mm^2) are painted on the lateral surfaces. The interelectrode spacing equals $L = 1$ mm. For the realization of the drift regime of signal excitation the external dc voltage was applied to the crystal.

First we have carried out standard measurements of the non-steady-state photo-EMF in $Bi_{12}SiO_{20}$ crystal. These experiments help us to find out necessary conditions for realization of the self excitation regime and analyze it. Frequency transfer functions of the non-steady-state photo-EMF are presented in Figure 5.4. The dependencies were measured for the zero external electric

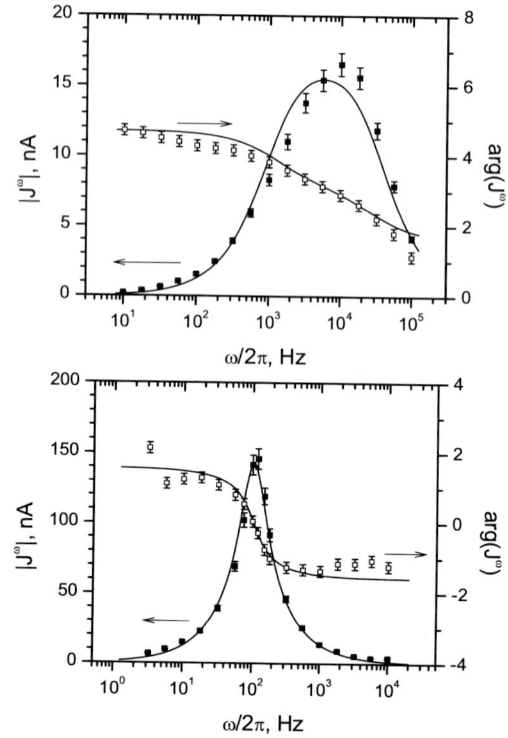

Figure 5.4. Amplitude and phase frequency transfer functions of the non-steady-state photo-EMF measured in $Bi_{12}SiO_{20}$ crystal at $E_0 = 0$ (top figure, $\Delta = 0.16$) and $E_0 = 10$ kV/cm (bottom figure, $\Delta = 0.32$).

field (diffusion regime of signal excitation) and rather large field of 10 kV/cm (drift regime). We have approximated them by Eqs. (5.7) and (5.19). The frequency of the maximum is one of the fitting parameters. The photocurrent is also characterized by the phase shift equal to π (diffusion regime) and 0 (drift regime) at this frequency. It was found to be $\omega_m/2\pi = 5.9$ kHz (for $E_0 = 0$) and $\omega_m/2\pi = 120$ Hz (for $E_0 = 10$ kV/cm). Namely these values are expected as the generation frequencies in the experiments with positive feedback.

When feedback loop with appropriated phase shift and gain factor is created the oscillations of the current in crystal and oscillations of the optical phase of the signal beam arise. The minimum gain factors are the following: $H \simeq -6.0 \times 10^3$ for the diffusion regime of the signal excitation and $H \simeq 1.4 \times 10^4$ for the drift one. Let us calculate the corresponding theoretical values using the material parameters from Ref. [45]: $sN_D = 1.06 \times 10^{20}$ J^{-1}m^{-1}, $\mu/(\gamma N_A) = 6.38 \times 10^{-11}$ m^{2}V^{-1}. Assuming the illuminated electrode area to be equal to $S = 3$ mm^2 (unexpanded beams, $E_0 = 0$) and $S = 9$ mm^2 (expanded beams, $E_0 = 10$ kV/cm) we obtain the following estimations: $H = -5.2 \times 10^3$ for the diffusion regime and $H = 4.5 \times 10^3$ for the drift one. One can note the good coincidence of the experimental and theoretical values for the former case and rather passable for the latter. We suppose the discrepancy obtained for case $E_0 \neq 0$ can be attributed to the voltage drop on the blocking contact reducing the electric field in the crystal's volume and space charge grating saturation effects.

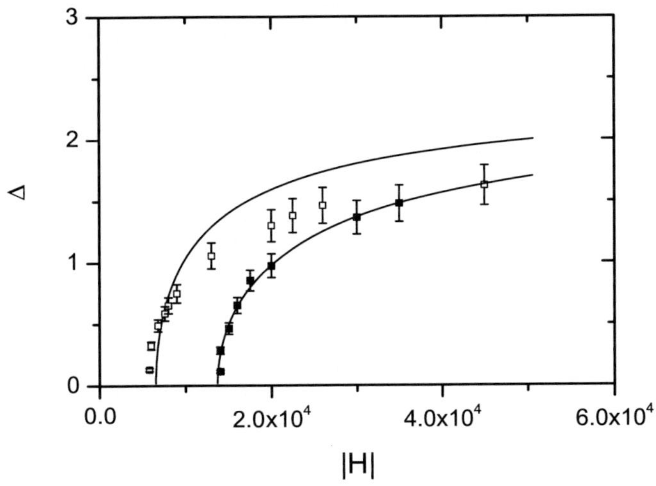

Figure 5.5. Dependencies of the phase oscillation amplitude on the gain factor of the amplifier measured in $Bi_{12}SiO_{20}$ crystal at $E_0 = 0$ (open squares) and $E_0 = 10$ kV/cm (filled squares).

We measured the dependencies of the optical phase oscillation amplitude on the gain factor of the voltage amplifier (Figure 5.5). One can note the following peculiarity. When the oscillations appear, they have rather small amplitude.

Besides we observed no hysteresis effect, i.e., oscillations arise and disappear at the same gain factor values. So we can state that the constructed optoelectronic generator is characterized by the "soft" regime of excitation. The experimental dependencies were approximated by Eq. (5.11).

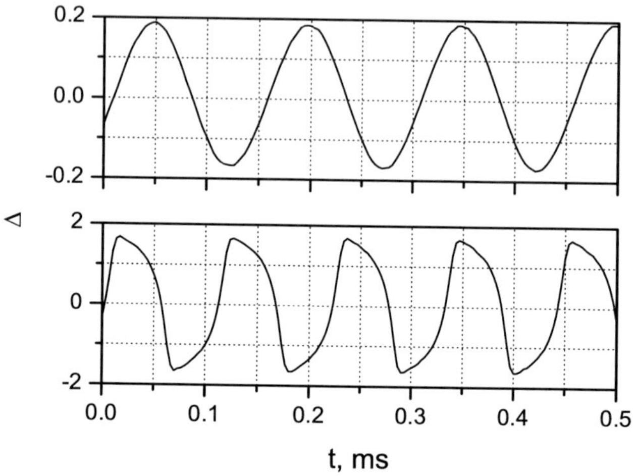

Figure 5.6. Light phase oscillations measured in $Bi_{12}SiO_{20}$ crystal for $E_0 = 0$ and for two gain factors of the amplifier: $H = -6.0 \times 10^3$ (top figure) and $H = -4.5 \times 10^4$ (bottom figure).

Let us consider the time dependencies of the oscillations (Figures 5.6, 5.7). They were measured for diffusion and drift regimes of the photo-EMF excitation and for two values of gain factor: one is near the excitation threshold and other is the maximum value where linear properties of the amplifier are guaranteed. One can note that oscillations are sinusoidal for the case of small gain factors. Their amplitude is rather small and frequency well corresponds to the frequency of current maximum in the experiments with forced excitation of the non-steady-state photo-EMF signal: $\omega_g/2\pi = 6.7$ kHz (for $E_0 = 0$, $H = 6.0 \times 10^3$) and $\omega_g/2\pi = 106$ Hz (for $E_0 = 10$ kV/cm, $H = 1.4 \times 10^4$). When gain factor increases the oscillations become nonharmonic and their frequency increases: $\omega_g/2\pi = 9.1$ kHz (for $E_0 = 0$, $H = 4.5 \times 10^4$) and $\omega_g/2\pi = 117$ Hz (for $E_0 = 10$ kV/cm, $H = 3.5 \times 10^4$). We noticed the important feature: the de-

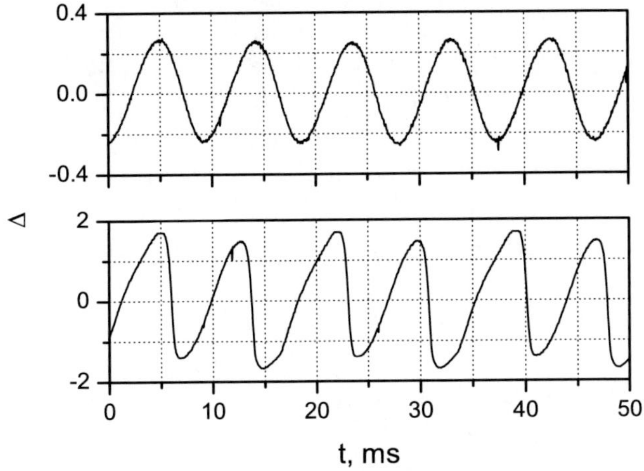

Figure 5.7. Light phase oscillations measured in $Bi_{12}SiO_{20}$ crystal at $E_0 = 10$ kV/cm and for two gain factors: $H = 1.4 \times 10^4$ (top figure) and $H = 3.5 \times 10^4$ (bottom figure).

viation of the generation frequency in the diffusion regime is larger than the deviation in the drift one for comparable increase of the gain factor. It is most probably due to the resonant nature of the signal excitation in the drift regime. It should be noted that the observed time dependencies demonstrated instability for certain gain factors. For low gain factors instability is associated with fluctuations in the experimental environment (instabilities of the laser power and external voltage, vibrations, etc.). In the drift regime and for large gain factors the subharmonics with frequencies $\omega_g/2, \omega_g/3$ appear and disappear during measurements (Figure 5.7). This feature can be associated with the effects of space charge wave instability described earlier [67, 229].

The possibility for measurements of photoelectric parameters was pointed out above. As follows from Eq. (5.10), the product $\tau\tau_M = 5.6 \times 10^{-10}$ s^2 can be evaluated directly from the oscillation frequency in the diffusion regime of recording ($E_0 = 0$). For the drift regime of the signal excitation ($E_0 = 10$ kV/cm) the oscillation frequency exactly equals to the eigenfrequency of the space charge waves, i.e., $\omega_{sc} = 670$ s^{-1}.

We have studied the new regime of the interferometer using detector based on the non-steady-state photo-EMF effect. Namely, oscillations of the optical phase are generated in the interferometer with the $Bi_{12}SiO_{20}$ detector and the optoelectronic feedback loop. The main parameters of such oscillations, i.e., amplitude and frequency, are defined mainly by the nonlinearity of the amplitude characteristic of the non-steady-state photo-EMF effect and photoelectric properties of the crystal and not by the experimental environment. This feature can be used for semiconductor testing applications. For example frequency of the oscillations excited in the drift regime of recording equals the eigenfrequency of the space charge waves – the fundamental mode of the space charge oscillations. Using this technique we have estimated this eigenfrequency along with another complex parameter – the product $\tau\tau_M$. Determination of the basic parameters like Maxwell relaxation time τ_M, photoconductivity relaxation time τ and carrier's mobility μ requires the wideband electronic amplifier with phase shift $\pm\pi/4$ and $\pi/2$.

The developed technique can also be applied for the detection of phase modulated optical signals, e.g., in laser ultrasonics. One can realize the regenerative principle of signal amplification in the considered interferometric scheme with feedback loop. In this case one of the interfering laser beams (Figure 5.1, a) should carry the information signal (e.g., from vibrating object), and dimensionless gain factor Ξ should be chosen slightly smaller than critical value ($\Xi < 2$). The amplitude of the detected photo-EMF signal can be increased several times with respect to the standard scheme [21] (as a rule, reachable magnification factor is defined by the stability of the components of the setup).

The analysis of the considered phenomena is based on the linear approach of the non-steady-state photo-EMF theory. It allowed us to describe the rise of the generation regime when the amplitude of the oscillations are small enough. However, some peculiarities observed in the experiments, e.g., frequency shift, nonharmonic form of the oscillations with large amplitude and subharmonics generation, point out to the necessity of the more detailed theoretical analysis of the effect.

2. Detection of Phase-Modulated Optical Signals

Let us proceed with the utilization of the laser vibrometers using GaAs adaptive photodetectors based on the effect of the non-steady-state photoelectromotive force. We shall also consider the influence of an external ac field on the signal-to-noise ratio for the nonresonant and resonant techniques of signal enhancement.

Adaptive photodetectors using GaAs crystals were tested at different wavelengths and the best sensitivity was obtained at (λ = 1064 nm). We have also performed measurements at the excitation wavelength of standard He-Ne laser (λ = 633 nm). For this illumination wavelength the absorption coefficient of semi-insulating GaAs is high (10^4 - 10^5 cm^{-1}) and the photocurrent signal is excited in a thin layer of material - about 1 micrometer. The signal beam is scattered by the rough surface and the reference beam is directly derived from the laser interfere inside the adaptive photodetector. The signal from adaptive photodetector is amplified and visualized on the screen of oscilloscope or spectrum analyzer. In our experiments at λ = 633 nm the carrier spatial frequency was chosen higher than the inverse size of the speckle spots, so an interference pattern with the spatial frequency 30 lines/mm was formed on every speckle spot. This spatial frequency is an optimal one for observation of non-steady-state photo-EMF in semi-insulating wafers of GaAs (it corresponds to the inverse diffusion length of photocarriers). The crystal was observed to be capable of measuring directly vibration frequencies, with a resolution of about 2 mHz, limited by the sensitivity of the spectrum analyzer. Our studies at λ = 1064 nm showed that it is possible to detect ultrasonic vibrations with the amplitude of 0.2 Å with a signal power of 20 mW and a bandwidth of 15 MHz.

Development and quantity production of gyroscopes, accelerometers in machine building, adaptive reflecting systems using MEMS needs precise and simple express-evaluation technique for the determination of frequency-amplitude characteristics of natural oscillations. MEMS is the integration of mechanical elements, sensors, actuators, and electronics on a common silicon substrate through microfabrication technology. While the electronics are fabricated using integrated circuit process sequences, the micromechanical components are fabricated using compatible micromachining processes that selectively etch away parts of the silicon wafer or add new structural layers to form the mechanical and electromechanical devices. Let us briefly discuss, for example, the operation of accelerometer consisting of immovable base and massive sensing head attached to the base via tips. During the process of acceleration measurement the head deviates and one can easily determine the capacity in the gap between the head and the contact attached to the base. The value of capacity change determines the acceleration value, the known spatial configuration of the sensing head allows determination of the acceleration vector change direction. Such system has its own resonant frequencies that should be known for proper calibration of the device. Theoretical calculation of such resonant frequencies does not give adequate precise data since it is practically impossible to take into account complicated geometry of MEMS and more important - specific features of its manufacturing ("MEMS errors"). It is important to point out that at present time two basic approaches for MEMS testing exists - the first one uses complicated and expensive systems of computer image processing. The second one is based on electrical methods of detection and need special design of the MEMS chips (i.e., initially the specially prepared chip is tested, after the MEMS withdrawal it is incorporated into the device which is time consuming operation). Note, that optical methods for MEMS testing are introduced as well. We have utilized adaptive interferometers for the MEMS testing. For MEMS excitation (frequency range 100 Hz-20 kHz) we used two approaches: the first one uses acoustic wave from standard loud speaker; the second one assumes contact vibration excitation. In the latter case the device was attached to piezoelectric actuator. The MEMS membranes were manufactured in the laboratory of Micro-Technologies of St.-Petersburg State Technical University with the size of 5×5 mm^2 and thickness of 500 microns. The technique enabled efficient direct conversion of high-frequency phase modulation optical wave reflected from vibrating MEMS structure into an electrical signal with concomitant setting of optimal operation point of the interferometer and suppression of amplitude laser noise. The fundamental resonant frequencies of MEMS structures were determined.

Let us analyze in more details the influence of an external ac field on the signal-to-noise ratio – the most important parameter of detectors. First, let us consider the nonresonant technique of signal enhancement [60]. In this case the crystal is placed in high frequency ac electric field: $\tau_M^{-1} \ll \Omega \ll \tau^{-1}$, where τ_M is the Maxwell relaxation time and τ is the lifetime of photoelectrons. The adaptive photodetector is illuminated by an oscillating interference pattern formed by two coherent light waves one of which is phase modulated with frequency ω (Figure 5.8). The nonresonant enhancement occurs under condition $\omega < \Omega$ and it is due to increase of the space charge field grating recorded in the

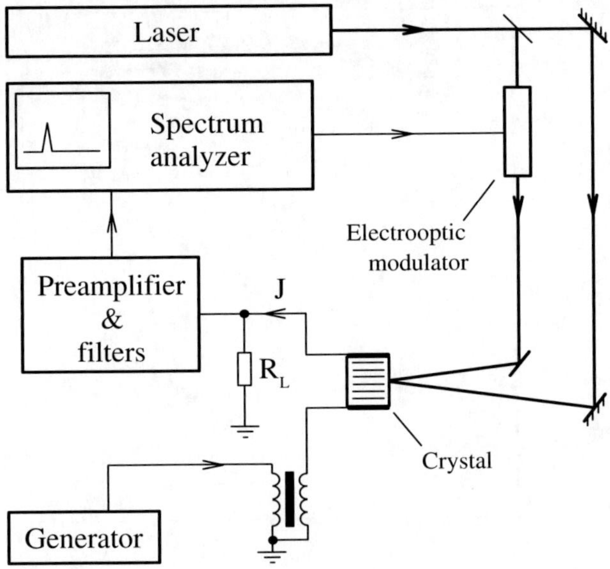

Figure 5.8. Experimental arrangement for the excitation and detection of the non-steady-state photo-EMF in the adaptive photodetector placed in an external ac field.

crystal volume [2, 60]. The physics of the signal enhancement was discussed in Ref. [60], experimentally the increase of the signal amplitude by 50 dB was obtained (see Figure 5.9). The expression for the amplitude of photocurrent excited in the presence of the sinusoidal electric field can be written as [60]:

$$J^{\omega} = \frac{-0.5Sm^2\Delta\sigma_0[E_D + E_0 K L_0/(1 + K^2 L_D^2)]i\omega\tau_M}{1 + i\omega\tau_M[1 + K^2 L_D^2 + K^2 L_0^2/(1 + K^2 L_D^2)]}. \tag{5.25}$$

Here S is the electrode area, m and K are the contrast and spatial frequency of the interference pattern, Δ is the amplitude of phase modulation of the signal beam, σ_0 is the average photoconductivity, $E_D = Kk_BT/e$ is the diffusion field, E_0 is the amplitude of external ac field, $L_D = \sqrt{(k_BT/e)\mu\tau}$ and $L_0 = \mu\tau E_0$ are the diffusion and drift lengths (k_B is the Boltzmann constant, T is the temperature, e is the electron charge and μ is the mobility). As follows from Eq. (5.25) the frequency transfer function of the ef-

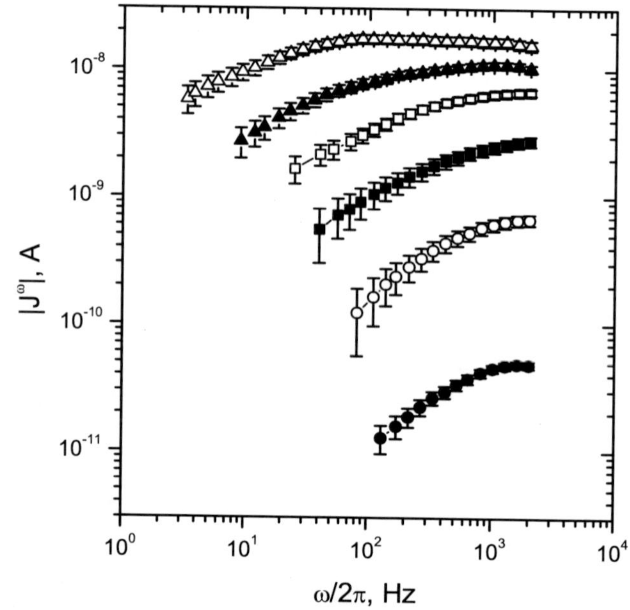

Figure 5.9. Frequency transfer functions of the photocurrent measured for $E_0 = 0$ (filled circles), 53 V/mm (open circles), 130 V/mm (filled squares), 210 V/mm (open squares), 410 V/mm (filled triangles), and 1000 V/mm (open triangles). $Bi_{12}SiO_{20}$, $\lambda = 442$ nm, $P_0 = 0.19$ mW, $m = 0.65$, $K = 5 \times 10^4$ m^{-1}, infrared illumination is switched on.

fect has the traditional form typical for the diffusion mechanism of photocurrent excitation [20], namely, the linear growth of the signal for low modulation frequencies and the frequency-independent region for high frequencies of phase modulation. These two regions are separated by the cut-off frequency $\omega_0 = \{\tau_M[1 + K^2L_D^2 + K^2L_0^2/(1 + K^2L_D^2)]\}^{-1}$. If we consider the frequency-independent region ($\omega \gg \omega_0$), which is more suitable for detection of phase modulated optical signals, and suppose the spatial frequency to be small ($K^2L_D^2 \ll 1$), then Eq. (5.25) can be simplified to the following one:

$$J^\omega(E_0) \simeq -\frac{Sm^2\Delta\sigma_0 E_D}{2}\frac{1 + E_0^2/E_L^2}{1 + E_0^2/E_M^2}, \tag{5.26}$$

here $E_L = (k_B T/e\mu\tau)^{1/2}$, $E_M = (K\mu\tau)^{-1}$ are the characteristic values of the electric field. Here we should note that Eqs. (5.25), (5.26) present the amplitude values of the photocurrent, the effective values are $\sqrt{2}$ times lower.

For the $\mu\tau$-product of $Bi_{12}SiO_{20}$ crystal ($\mu\tau \simeq 10^{-10}$ m^2/V) and for low spatial frequency ($K \simeq 10^5$ m^{-1} $< L_D^{-1}$) we have $E_L \simeq 0.16$ kV/cm and $E_M \simeq 1.0$ kV/cm. For such relation between the characteristic fields (i.e., $E_L < E_M$) one can see from Eq. (5.26), that there is a region ($E_L < E_0 < E_M$) on dependence $J^\omega(E_0)$ where the photocurrent amplitude grows as E_0^2. This behavior was also observed experimentally [60]. For such fast growth of the signal amplitude one should also expect the increase of the signal-to-noise ratio. Let us calculate corresponding noise and signal amplitudes as well as signal-to-noise ratio.

Wide-gap material has much higher resistance than the load resistor and input circuits ($R_{cr} \gg R_L$). This means that for diffusion mechanism of operation (no external field $E_0 = 0$) the noise of the photo-EMF-based adaptive detector is defined by the thermal noise of the load resistor R_L:

$$J_{th} = \sqrt{4k_B T R_L^{-1} \Delta f}, \tag{5.27}$$

where Δf is the detection bandwidth. Dividing Eq. (5.26) by Eq. (5.27) we obtain the signal-to-noise ratio for the case of zero external field. For low spatial frequencies ($K^2 L_D^2 \ll 1$) it equals

$$(S/N)_0 = \frac{Sm^2\Delta}{4\sqrt{2}}\sigma_0 E_D \sqrt{\frac{R_L}{k_B T \Delta f}}. \tag{5.28}$$

For the matched load ($R_{cr} \sim R_L$) thermal noise of load resistor and generation-recombination noise become comparable and Eq. (5.28) slightly changes [230].

Let us consider the case on nonzero applied ac fields. In practice, external fields of the order of $E_L \sim 0.1$ kV/cm increase the generation-recombination noise in adaptive detector making thermal noise contribution to be negligible. The nature of the generation-recombination noise as well as the procedure for its calculation is well-known (see, for example, Ref. [62]), for this reason here we shall use the expression presented in Ref. [230]:

$$J_{g/r} = 2e\mu\tau E_0 \sqrt{S g_0 \Delta f / L}. \tag{5.29}$$

Here g_0 is the average generation rate of photoelectrons, L is the interelectrode spacing, condition $\omega, \Omega \ll \tau^{-1}$ is assumed to be fulfilled. The fact that photocurrent amplitude [Eq. (5.26)] grows faster than noise [Eq. (5.27)] can be used for the increase of the signal-to-noise ratio of the device. This value reaches its maximum for the optimal electric field $E_0 = E_M$:

$$(S/N)_{E_M} = \frac{m^2\Delta}{8\sqrt{2}}\sqrt{\frac{g_0 SL}{\Delta f}}. \tag{5.30}$$

The magnification of the signal-to-noise ratio can be characterized by the following relation:

$$\frac{(S/N)_{E_M}}{(S/N)_0} = \frac{1}{2KL_D}\sqrt{\frac{R_{cr}}{R_L}}. \tag{5.31}$$

So the application of the external field can be useful for the situation when the spatial frequency, crystal's photoconductivity and photocarriers' diffusion length are small. Let us estimate this factor for the parameters of our experiment (see Ref. [60]): $K = 5 \times 10^4$ m^{-1}, $L_D = 1.7$ μm, $R_{cr} \sim 20$ MΩ, $R_L = 15$ kΩ. After substitution we obtain $(S/N)_{E_M}/(S/N)_0 \simeq 200 = 46$ dB. In reality this ratio did not exceed 10 dB, what is most probably associated with the additional current noise due to amplitude noise of the laser used in our experiments presented in Ref. [60].

Let us proceed with the analysis of the signal-to-noise ratio for the case of resonant excitation of non-steady-state photocurrent [155]. The giant signal enhancement in the high frequency region ($\omega \gg \tau^{-1}$) is the most attractive feature of the signal generation in an alternating electric field ($\tau_M^{-1} \ll \Omega \ll \tau^{-1}$). The enhancement of the signal amplitude reaches ~ 70 dB for low spatial frequencies of the interference pattern (Figure 5.10). For practical applications it is more convenient to calculate and compare the signal-to-noise ratio for the zero and ac fields. Let us estimate this value for the resonant frequency ($\omega_r = K\mu E_0 \gg \tau^{-1}, \tau_M^{-1}$), low spatial frequency of the interference pattern ($K^2L_D^2 \ll 1$), and low load resistance ($R_L \ll |Z_{cr}|$, Z_{cr} is the crystal's impedance).

For the discussed frequency region the signal amplitude can be expressed as (see Ref. [155]):

$$J^\omega = -\frac{Sm^2\Delta}{2}\sigma_0\left(E_D + E_0\frac{KL_0}{1+K^2L_D^2}\right)$$

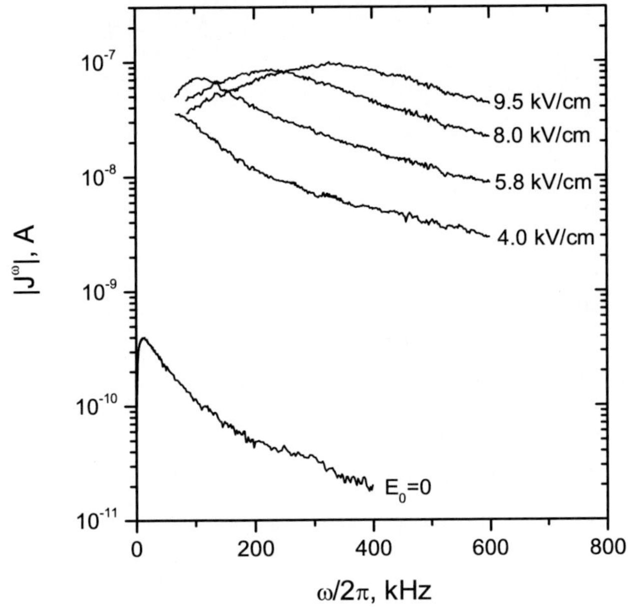

Figure 5.10. Frequency transfer functions of the holographic current measured in photorefractive $Bi_{12}SiO_{20}$ for different effective values of the sinusoidal electric field: $E_0 = 0$, 4.0, 5.8, 8.0, 9.5 kV/cm ($\lambda = 532$ nm, $I_0 = 2.2$ W/cm^2, $K = 500$ cm^{-1}, $\Omega/2\pi = 6$ kHz).

$$\times \frac{1 + K^2L_D^2 + i\omega\tau}{(1 + K^2L_D^2 + i\omega\tau)^2 + K^2L_0^2}. \tag{5.32}$$

As seen from the expression, the signal amplitude grows as E_0^2, so we can also expect the improvement of the signal-to-noise ratio of adaptive photodetector.

As it was mentioned above, the thermal noise of the load resistor [Eq. (5.27)] is dominant for the diffusion regime of the photocurrent generation ($E_0 = 0$). The signal-to-noise ratio $(S/N)_0$ for the high-frequency region ($\omega > \tau^{-1}$) equals:

$$(S/N)_0 = \frac{Sm^2\Delta}{4\sqrt{2}} \frac{\sigma_0 E_D}{\omega\tau} \sqrt{\frac{R_L}{k_B T \Delta f}}. \tag{5.33}$$

It is evidently smaller than that described by Eq. (5.28) because of the fast signal decay for modulation frequencies $\omega > \tau^{-1}$.

For the applied external ac voltage the generation-recombination noise is noticeable. For the high frequency region it is described by the following expression [62]:

$$J_{g/r} = 2e\mu\tau E_0 \sqrt{\frac{Sg_0\Delta f}{L(1+\omega^2\tau^2)}}. \tag{5.34}$$

Since the amplitude of the generation-recombination noise decays as ω^{-1} at $\omega > \tau^{-1}$ we cannot neglect contribution of the thermal noise of the load resistor when calculating the signal-to-noise ratio. Finally, the signal-to-noise ratio at the resonant frequency $\omega = \omega_r$ equals

$$(S/N)_{E_0} = \frac{Sm^2\Delta\sigma_0 E_0 K L_0}{8\sqrt{2}\sqrt{k_B T R_L^{-1}\Delta f + Se^2 g_0 \Delta f/LK^2}}. \tag{5.35}$$

Dividing Eq. (5.35) by Eq. (5.33) we obtain the magnification of the signal-to-noise ratio at resonant frequency ($\omega = \omega_r$):

$$\frac{(S/N)_{E_0}}{(S/N)_0} = \frac{E_0 K^2 L_0^2}{2\sqrt{E_D^2 + k_B T g_0 R_L S/L}} \tag{5.36}$$

This gain factor is very large: for the sensible experimental conditions and crystal parameters ($E_0 = 10$ kV/cm, $K = 500$ cm^{-1}, $L_0 = 0.02$ cm, $L = 0.1$ cm, $S = 10^{-2}$ cm^2, $g_0 = 10^{19}$ cm^{-3}s^{-1}, $R_L = 10^4$ Ω) it reaches 90 dB. In practice, however, some factors can reduce this striking value: nonlinear recording of the space charge grating, trap saturation, etc. In our experiments (Ref. [155]) we observed the 70 dB increase of the signal while noise increases within 3 dB, so the gain factor was about 67 dB.

In this Chapter we have discussed operation of adaptive interferometers and analyzed the influence of an external ac electric field on the signal-to-noise ratio of the devices. It is shown that signal-to-noise ratio can be sufficiently increased via application of external field. The gain factor calculated for parameters of n-type photorefractive $Bi_{12}SiO_{20}$ crystal can reach 46 dB for low excitation frequencies and 90 dB for higher frequencies. We believe that the considered techniques of signal-to-noise improvement can be applied for adaptive photodetectors based on other semiconductor crystals (GaAs, GaN, ZnSe, etc.).

3. Visualization of Oscillatory Stresses in Transparent Media using a β-Ga_2O_3 Adaptive Detector

The high sensitivity, operation with complicated wavefronts and remote sensing have assured the success of the holographic interferometry in the investigation of the gas and liquid streams and in the nondestructive testing of solids [231]. The utilization of the photorefractive crystals adds the possibility of the real-time processing and visualization of the holograms [1, 2]. The reversibility of the hologram formation in the photorefractive media makes the interferometer to be adaptive, i.e., able to maintain the optimal operating point in the presence of the slow phase drift caused by thermal drift of the interferometric setup, air flows, etc. There are some modern applications of the interferometry, where detection of the response to an oscillating but not stationary force is preferable or even necessary [204, 205, 232–238], and the adaptive interferometers using photorefractive crystals perfectly match this tasks.

The most of mentioned properties of the photorefractive effect is also inherent in the holography-related phenomenon called non-steady-state photoelectromotive force (photo-EMF). This effect reveals itself as an alternating electric current arising in a semiconductor illuminated by an oscillating interference pattern. The electric current appears due to spatial shifts of the space-charge and photoconductivity distributions in the crystal volume. The excitation of the non-steady-state photo-EMF signal involves processes of carrier generation, diffusion and drift in an electric field, recombination to local centers. This complex nature of the effect makes it a powerful tool for determination of photoelectric parameters such as type and value of photoconductivity, carrier lifetime, mobility and diffusion length, density of recombination centers, etc. [20, 32, 100, 132–134, 154, 173, 174]. Besides the tasks of material characterization the non-steady-state photo-EMF is used for the detection of phase- and frequency-modulated optical signals [21, 130, 136, 138, 139, 141, 142, 230, 239–241].

The detection of phase-modulated optical signals using adaptive detectors is usually carried out using the vibrometric scheme, where the light phase incursion is created by the reflection from an oscillating object. Such an additional phase can obviously appear in the light beam propagating in a stressed transparent material as well [231]. The most of the developed interferometric techniques deals with the stationary stress fields, while the distributions of time-varying stresses are also of interest, since they can greatly differ from the stationary ones. So, the main aim of this section is to apply the non-steady-state photo-EMF technique to the detection of phase-modulated optical signals produced by the oscillating stresses in the transparent material [242]. The experiments are performed with the adaptive detector made of β-Ga_2O_3 crystal and with the conventional glass plate. The crystal of gallium oxide is considered as one of the most promising materials for production of solar-blind photodetectors [180], where it can replace silicon and GaAs detectors with their concomitant UV filters. The wavelength shortening requires not only novel detectors but also opens up the need for better instrumentation, including better windows, gratings, filters, as well as the methods for their non-destructive testing [243]. So, the combination of the new wide-bandgap material and modified technique seems to be of vital importance. After experiments with the β-Ga_2O_3 detector we simulate the phase response induced by the oscillating force via the piezooptic effect and compare it with the experimental data.

The experiments with the detection of oscillatory stresses are carried out with the arrangement (Fig. 5.11), which is almost the same as that described in Section 3.7. One of the beams passes through the transparent glass plate PM-15 (Fig. 5.11). The plate is made of soviet glass K8 and has the diameter of 30.3 mm and thickness of 15.87 mm. The front and back surfaces have the optical quality. The lateral (cylindric) surface has two diametrically arranged 6.6 mm flats, which were used for clamping. The oscillatory stresses are created in this plate by the piezoelectric transducer. The electro-optic modulator is used for the calibration of the β-Ga_2O_3 detector. Being switched on it introduces phase modulation with amplitude $\Delta_c = 0.61$ and frequency ω into the other arm of the interferometer.

We have measured the frequency transfer function of the non-steady-state photo-EMF amplitude in our β-Ga_2O_3 detector, i.e., dependence $J^\omega = J^\omega(\Delta_c, \omega)$ with the constant Δ_c (Fig. 5.12). The dependence contains the growth of the signal at low frequencies ω followed by the frequency independent part. The cut-off frequency is of 13 Hz, and it corresponds to the following values of the Maxwell relaxation time and photoconductivity: $\tau_M = 10$ ms, $\sigma_0 = 1.0 \times 10^{-10}\ \Omega^{-1}\text{cm}^{-1}$ [175, 242]. The dependence of the signal amplitude on the spatial frequency of the interference pattern is another characteristic usually measured in the non-steady-state photo-EMF experiments (Fig. 5.12).

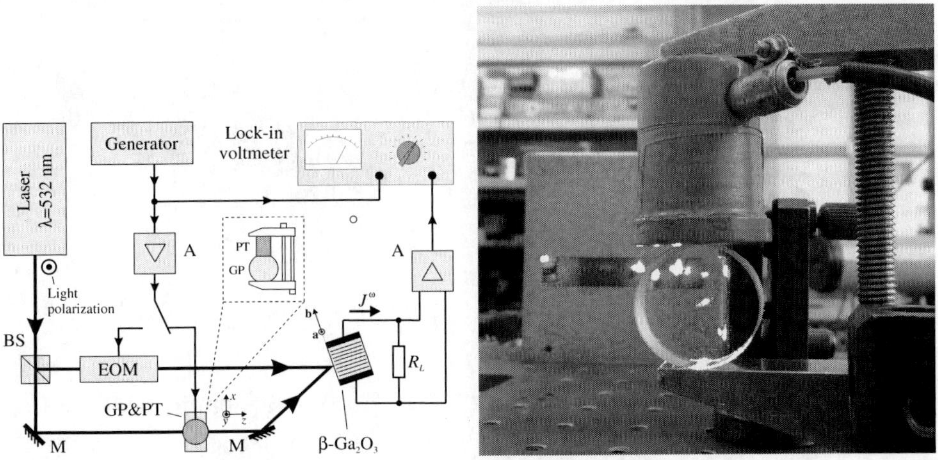

Figure 5.11. Experimental setup for the excitation and detection of oscillatory stresses in the glass plate (left). EOM is the electrooptic modulator, BS is the beamsplitter, M is the mirror, A is the amplifier, GP&PT are the glass plate and piezoelectric transducer. The orientation of the coordinate system used for the simulation is shown near the glass plate. Photograph of the glass plate and piezoelectric transducer vised in the locksmith clamp (right). The illuminated β-Ga_2O_3 detector placed in the metallic shield is in the background.

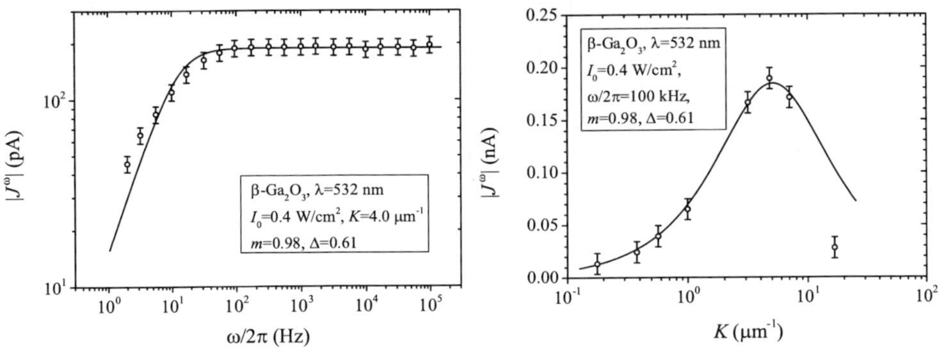

Figure 5.12. Frequency dependence of the non-steady-state photo-EMF in the β-Ga_2O_3 adaptive detector (left). Dependence of the non-steady-state photo-EMF amplitude on the spatial frequency of the interference pattern (right). The solid lines show the approximation by Eq. (3.67).

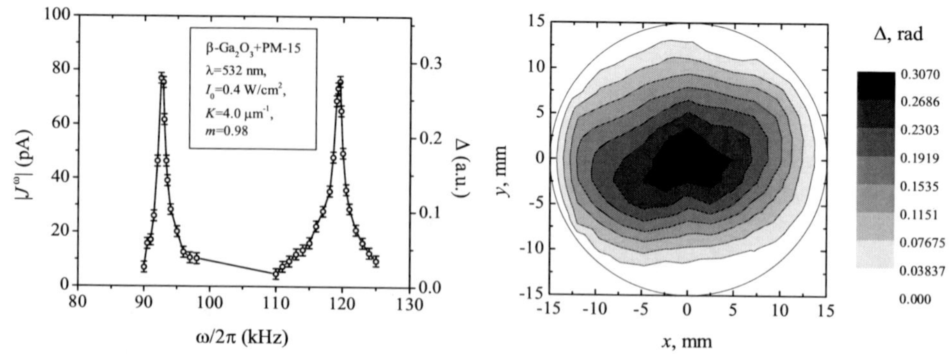

Figure 5.13. Frequency dependence of the non-steady-state photo-EMF excited in the β-Ga_2O_3 adaptive detector by the light passed through the oscillating PM-15 glass plate (left). The right axis is for the calculated amplitude of phase modulation. Distribution of the phase modulation amplitude across the glass plate surface (right). The plot corresponds to the photo in Fig. 5.11, i.e., the light propagates from the reader.

The spatial frequency $K = 4.0\ \mu\text{m}^{-1}$ chosen for the next experiments with the glass plate is very close to its optimum value.

After calibration of the adaptive photodetector one can measure the frequency response of a real oscillating object. The frequency dependence of the non-steady-state photo-EMF signal excited by the oscillations of the PM-15 glass plate is presented in Fig. 5.13. In general such a dependence contains both the “object” and “detector” contributions: $J^\omega = J^\omega(\Delta(\omega), \omega)$. In order to evaluate the amplitude of the phase modulation $\Delta(\omega)$ resulted from mechanical oscillations of the object one has to solve the following nonlinear equation:

$$\frac{\mathrm{J}_0(\Delta(\omega))\,\mathrm{J}_1(\Delta(\omega))}{\mathrm{J}_0(\Delta_c)\,\mathrm{J}_1(\Delta_c)} = \frac{J^\omega(\Delta(\omega), \omega)}{J^\omega(\Delta_c, \omega)} \tag{5.37}$$

The amplitude of phase modulation produced by the glass plate is small enough ($\Delta < 0.3$), and the studied frequency range $90 - 125$ kHz corresponds to the plateau in the frequency transfer function of the β-Ga_2O_3 adaptive detector. So this equation can be linearized, and the desired frequency dependence $\Delta(\omega)$ can be calculated just by the multiplication of the measured dependence $J^\omega(\Delta(\omega), \omega)$ by a certain coefficient or by adding another axis in the plot, as it was done in Fig. 5.13. The resonant behavior of the signal J^ω and calculated amplitude Δ is associated with the mechanical resonances of the system including the glass plate, piezoelectric transducer and clamp. The increase of the applied constant tension to the plate and transducer slightly increases the resonant frequencies.

Since the glass plate has the rather complicated geometry one can expect that the mechanical stresses, deformations and resulting amplitude of phase modulation non-uniformly distribute across the plate surface. We have measured this distribution by moving the clamp with the glass plate and transducer in the directions perpendicular to the light beam. The measurements are performed at $\omega/2\pi = 93.5$ kHz, i.e., at the first resonant frequency. As seen, the distribution has the nearly radial symmetry with the maximum at the center of the plate. The slight counterclockwise slope of the distribution can be associated with the non-uniformity of the applied periodical tension across the lateral surface's flats.

The calculation procedure for the piezooptic effect is well known from the literature [244]. It includes four steps: (1) the definition of the mechanical stress tensor σ in the crystal-physical coordinate system, (2) calculation of dielectric impermeability tensor $\xi_{ij} = \pi_{ijkl}\sigma_{kl}$ in the crystal-physical coordinate system, (3) transformation of tensor ξ to the principal coordinate system which is defined by direction of light propagation, (4) derivation of the refractive indexes $n_{1,2}$ and corresponding components of the electric induction vector $\mathbf{D}$. The glass is isotropic, and this greatly simplifies the problem. An arbitrary coordinate system can be considered as the crystal-physical. In particular, we can choose the system, where z-axis coincide with the direction of the light propagation, y-axis is parallel to the initial polarization plane of the light. This means that such a crystal-physical coordinate system is also a special one, and step (3) is not necessary. The tensor of piezooptic coefficients π of the isotropic media is defined by the only two independent values π_{11} and π_{12} (here and below we use both the full and reduced notation of tensors).

First let us calculate the oscillating mechanical stresses in the cylindric glass plate with the above mentioned parameters. The analytic approach for such a problem requires solution of motion equation for elastic medium [245]. Since the cylinder has the two flats, and its height is finite, the analytic approach seems too difficult. Besides, the obtained solution would be valid only for this geometry, which was chosen rather arbitrarily. So, the numerical solution is much more reasonable, and we use the structural mechanics module of COMSOL Multiphysics software to cope with this problem. The geometry of the problem is the same as that used in the experiment (Fig. 5.11). The bottom flat of the cylinder is fixed, the top flat oscillates with amplitude $Y = 1.5$ nm and frequency $\omega/2\pi = 104$ kHz. These two parameters are settled so that the final calculated distribution of the optical phase amplitude was close to the experimental one (Figs. 5.13, 5.15). The density $\rho = 2.51$ g/cm^3, Young's modulus $E = 82$ GPa and Poisson's ratio $\nu = 0.206$ of glass BK7 (analog of soviet glass K8) are used as the necessary material parameters.

Figure 5.14 presents the diagonal components of the stress tensor σ calculated for the cross section at $z = d/2$, i.e., at half hight of the cylinder. The analogous distributions of these values at different z including cylinder faces were calculated. They slightly differ from those presented in Fig. 5.14, so in the further analysis we can consider stresses σ_{11}, σ_{22} and σ_{33} as functions of coordinates x and y but not z. Stress σ_{11} is noticeably higher than σ_{22} and σ_{33} in the central area of the plate, but we do not neglect none of them until piezooptic and elastooptic effects are considered. The non-diagonal components of the tensor are not shown, since they do not contribute to the optical phase incursion, as it is shown below.

Now let us estimate the amplitude of phase modulation of the optical wave propagating through the stressed glass plate. The complex amplitude of the electric induction vector is the following at the input of the plate:

$$\mathbf{D}(x, y, 0) = D_0 \mathbf{e_y}, \tag{5.38}$$

where D_0 is the coordinate-independent value and $\mathbf{e_y}$ is the unit vector of the chosen coordinate system. On passing through the birefringent material with thickness z this vector equals

$$\begin{aligned} \mathbf{D}(x, y, z) &= \mathbf{D_1}(x, y) \exp\left[i\frac{2\pi}{\lambda} n_1(x, y) z\right] \\ &+ \mathbf{D_2}(x, y) \exp\left[i\frac{2\pi}{\lambda} n_2(x, y) z\right], \end{aligned} \tag{5.39}$$

where $n_1(x, y)$ and $n_2(x, y)$ are the refractive indexes of the eigenmodes in the birefrigent medium, and $\mathbf{D_1}(x, y)$ and $\mathbf{D_2}(x, y)$ are the corresponding induction vector components polarized along the main axes of the optical indicatrix

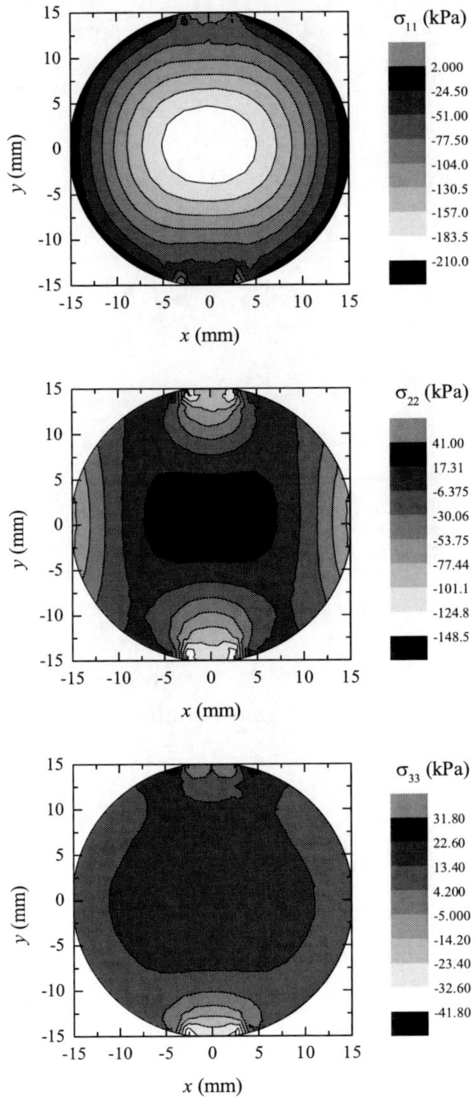

Figure 5.14. Calculated spatial distributions of the diagonal components σ_{11}, σ_{22}, σ_{33} of the stress tensor.

cross section. Amplitudes $\mathbf{D_1}(x, y)$ and $\mathbf{D_2}(x, y)$ are defined by the amplitude of the incident light D_0 and by the angle $\varphi(x, y)$ between vectors $\mathbf{D_1}(x, y)$ and $\mathbf{e_x}$:

$$\begin{aligned} \mathbf{D_1}(x, y) &= D_0 \sin\varphi(x, y)[\mathbf{e_x} \cos\varphi(x, y) + \mathbf{e_y} \sin\varphi(x, y)], \quad (5.40) \\ \mathbf{D_2}(x, y) &= D_0 \cos\varphi(x, y)[-\mathbf{e_x} \sin\varphi(x, y) + \mathbf{e_y} \cos\varphi(x, y)]. \quad (5.41) \end{aligned}$$

The expressions for values $n_{1,2}(x, y)$ and $\varphi(x, y)$ are known from the literature:

$$\begin{aligned} n_{1,2} &= n_0 - \frac{n_0^3}{4}\left[\xi_{11} + \xi_{22} \pm \sqrt{(\xi_{11} - \xi_{22})^2 + (2\xi_{12})^2}\right], \quad (5.42) \\ \varphi &= \frac{3\xi_{12}}{\xi_{11} - \xi_{22} + \sqrt{(\xi_{11} - \xi_{22})^2 + (2\xi_{12})^2}}, \quad (5.43) \end{aligned}$$

where $n_0 = 1.52$ is the refractive index of the unstressed material.

The non-steady-state photo-EMF signal is excited by the interference pattern, which is created by the light waves with the identical polarization states. The reference beam do not pass through the glass plate, it is polarized strictly along y-axis. In this sense only the y-component of the light passed through the plate is of interest:

$$\begin{aligned} [\mathbf{D}(x, y, z)]_y &= D_0 \sin^2\varphi(x, y) \exp\left[i\frac{2\pi}{\lambda}n_1(x, y)z\right] \\ &\quad + D_0 \cos^2\varphi(x, y) \exp\left[i\frac{2\pi}{\lambda}n_2(x, y)z\right]. \quad (5.44) \end{aligned}$$

On the other hand this component can be presented as follows:

$$[\mathbf{D}(x, y, z)]_y = D'(x, y) \exp\left[i\frac{2\pi}{\lambda}n_0 z + i\Delta_s(x, y)\right], \quad (5.45)$$

where D' is some amplitude and Δ_s is the desired light phase increment caused by the mechanical stress.

The substitution of Eqs. (5.42) and (5.43) into Eq.(5.44) provides the way to derive phase Δ_s. The problem can be simplified in the case of small stresses and corresponding light phase increments. After some routine calculations we have $D'(x,y) \simeq D_0$ and

$$\Delta_s(x,y) = -\frac{\pi n_0^3 z}{\lambda}\xi_{22}(x,y). \tag{5.46}$$

We can rewrite this expression using the tensor of piezooptic coefficients π_{ijkl} and distribution of the stresses $\sigma_{kl}(x,y)$:

$$\begin{aligned}\Delta_s(x,y) &= -\frac{\pi n_0^3 z}{\lambda}[\pi_{2222}\sigma_{22} + \pi_{2211}\sigma_{11} + \pi_{2233}\sigma_{33}] \\ &= -\frac{\pi n_0^3 z}{\lambda}[\pi_{11}\sigma_2 + \pi_{12}(\sigma_1 + \sigma_3)].\end{aligned} \tag{5.47}$$

The non-diagonal components of the stress tensor (σ_{12}, σ_{13}, σ_{23}) do not contribute to the phase increment since the corresponding components of tensor π_{ijkl} vanish in the isotropic medium.

The mechanical stresses modulate optical phase not only via changes of the refractive index but also via sample elongation δz in the direction of the light propagation:

$$\Delta_d = \frac{2\pi}{\lambda}n_0\delta z = \frac{2\pi}{\lambda}n_0\epsilon_{33}z. \tag{5.48}$$

Here ϵ_{33} is the component of the strain tensor. The Hooke's law connects this value with the stress components: $\epsilon_{33} = E^{-1}[\sigma_{33} - \nu(\sigma_{11} + \sigma_{22})]$. Finally we obtain the total phase increment, which is the sum of the "stress" and "deformation" parts:

$$\begin{aligned}\Delta(x,y) &= -\frac{2\pi}{\lambda}n_0 z\left[\left(\frac{n_0^2\pi_{12}}{2} + \frac{\nu}{E}\right)\sigma_1(x,y)\right. \\ &+ \left(\frac{n_0^2\pi_{11}}{2} + \frac{\nu}{E}\right)\sigma_2(x,y) \\ &+ \left.\left(\frac{n_0^2\pi_{12}}{2} - \frac{1}{E}\right)\sigma_3(x,y)\right].\end{aligned} \tag{5.49}$$

The calculated distributions of the phase components Δ_s, Δ_d as well as the total phase Δ are shown in Fig. 5.15. The values of piezooptic coefficients $\pi_{11} = 0.299 \times 10^{-12}$ Pa^{-1} and $\pi_{12} = 1.890 \times 10^{-12}$ Pa^{-1} are taken from

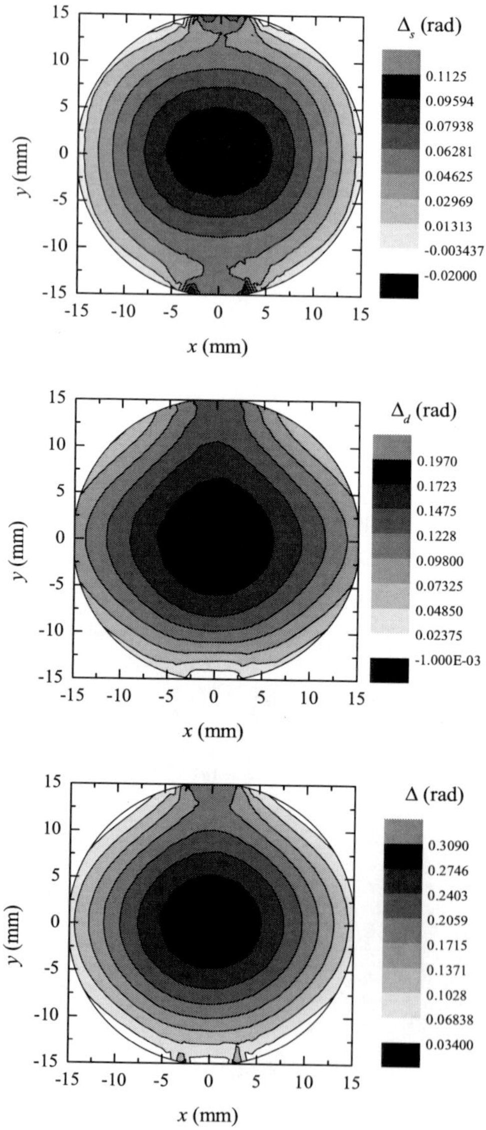

Figure 5.15. Calculated spatial distributions of the phase components Δ_s, Δ_d and the total phase Δ.

Ref. [246]. As seen, the "stress" and "deformation" components have comparable amplitudes, and none of them should be neglected in the total phase. The distributions are symmetric with respect of y-axis as expected. The maximum of the phase modulation amplitude is located in the vicinity of the plate's center, as it was observed in the experiment (Fig. 5.13). Distributions Δ_s, Δ are slightly sharpened in the top half, which is also the feature of the experimental dependence.

We have used the numerical solution of the elasticity equations for the plate, which has the rather complex lateral surface, namely, it is a cylindric one with two flats. However, there are analytic approaches developed for the ideal finite circular rod with stress-free boundaries [247–249]. These approaches allow calculation of the natural frequencies and corresponding mode shapes. The lowest dimensionless natural frequency corresponding to the "even" axisymmetric oscillation is estimated as 3.2 [247], which is equivalent to 120 kHz for the used glass (here we assumed, that the natural frequency weakly depends on the Poisson's ratio ν [248]). The frequency of the applied force is lower than this value, so the solution should be determined mainly by the lowest modes. This solution is presented as series [247, 249]. The terms in these series are expressed via the the linear combination of the radial functions $\mathrm{J}_0(\alpha_n r)$ and $(\alpha_n r)^{-1}\mathrm{J}_1(\alpha_n r)$ multiplied by the trigonometric function of coordinate z, where $\mathrm{J}_n(x)$ is the Bessel function of the first kind of the n-th order. Parameters α_n possesses several values, in particular, they are the roots of equation $\mathrm{J}_1(\alpha_n R) = 0$, where R is the radius of the cylinder. The mentioned radial functions are very similar to the bell-shaped distributions measured in the experiment and simulated numerically in this work. One can note, that the value of the natural frequency noticeably differs from the frequency used in the experiment (93.5 kHz) and simulation (104 kHz). We attribute the mismatch of these three frequencies to the complexity of our mechanical system, which include the glass plate, piezoelectric transducer and clamp with their own resonant frequencies. The mechanical contacts of these elements were not controlled and optimized, which can also introduce additional errors.

In this work we use the perspective semiconductor β-Ga_2O_3 as a material for adaptive detector of phase-modulated optical signals. This choice is not evident: the non-steady-state photo-EMF is excited by the light with the wavelength $\lambda = 532$ nm and photon energy $h\nu = 2.33$ eV, while the crystal's band gap is of 4.8 eV. Other materials with narrower band gaps, such as commonly used GaAs crystal, would have the higher photoconductivity and resulting current amplitude at this wavelength. However, the photoconductivity of the studied Ga_2O_3 crystal is monopolar, and the photo-EMF signal behavior is well described by Eq. (3.67). The bipolar conductivity of GaAs crystals in the visible and even near infrared ranges introduces some features, such as the partial compensation of the space-charge gratings, change of the signal sign at certain spatial frequency, appearance of the additional plateau and cut-off frequency due to the slow relaxation of carriers [86, 88], which can noticeably complicate the interpretation of obtained results. Besides, the sensors made of centrosymmetric crystals (β-Ga_2O_3) do not require additional damping of the vibrations resulting from the piezoelectric effect typical for non-centrosymmetric ones (GaAs).

Let us estimate the sensitivity of this detector operating in the linear regime ($m, \Delta \ll 1$) at the chosen spatial and phase modulation frequencies ($K = 4.0$ μm^{-1}, $\omega/2\pi = 100$ kHz). It is defined as the signal amplitude normalized to the power of the signal beam and amplitude of phase modulation. This provides the following estimate $S^{\omega} = J^{\omega}/(P_s\Delta) = 1.5 \times 10^{-7}$ A$\cdot$W^{-1}rad^{-1}. The noises in our measuring system are mainly determined by the thermal noise of the load resistor $R_L = 300$ kΩ, so the minimal detectable amplitude of phase modulation expresses as $\Delta_{min} = (4k_BTR_L^{-1}\delta f)^{1/2}/(S^{\omega}P_s)$, where δf is the detection bandwidth. For the sensible values of the power of the signal beam $P_s = 1$ mW and detection bandwidth $\delta f = 1$ Hz it equals $\Delta_{min} = 1.6 \times 10^{-3}$ rad. Since the piezooptic coefficients π_{ij} are of the same order of magnitude and the Poisson's ratio ν is of order of unity, we can roughly estimate the minimal detectable amplitude of the stress oscillation using Eq. (5.49):

$$\sigma_{min} \sim \left[\frac{2\pi}{\lambda} n_0 z \left(\frac{n_0^2 \pi_{ij}}{2} + \frac{\nu}{E}\right)\right]^{-1} \Delta_{min} = (1-2) \times 10^3 \text{ Pa}. \qquad (5.50)$$

Here we should note, that some experimental conditions could be changed to increase the signal-to-noise ratio, which is of 56 dB in Fig. 5.12. The theoretical limit of this ratio [230]

$$(S/N)_{max} = \frac{1}{8\sqrt{2}} \left(\frac{g_0 V}{\delta f}\right)^{1/2} \qquad (5.51)$$

is achieved at $m, \Delta, KL_D = 1$, $\omega \gg \omega_1$ and $Z_L = Z_{cr}^*$, where g_0 is the average generation rate of electrons, V is the crystal volume, Z_L and Z_{cr} are the

the load and crystal impedances. The used parameters m, Δ, K and ω are very close to their optimal values, their modifications can increase the signal by the factor of 1.4 only. The impedance matching seems the most challenging technical problem. The crystal resistance and capacitance are of 80 GΩ and 0.13 pF, which are calculated using the geometric dimensions and material parameters σ_0 and ϵ_{22}. The transimpedance amplifier allows the compensation of the source capacitance in the wide frequency region, but its design seems very artistic for such a large impedance. The crystal capacitance is comparable with the parasitic ones of the electronic circuit, and any small mismatch can result in the nonuniformity of the transfer function and even instability of the amplifier. That is why we prefer the commonly used short-circuit regime of the signal detection, which insures the flatness of the frequency transfer function up to $1/2\pi R_L C_L \sim 1$ MHz ($C_L \sim 0.5$ pF). However, if the all mentioned optimizations were performed, the maximum signal-to-noise ratio defined by Eq. (5.51) would reach striking 137 dB for $\delta f = 1$ Hz (here we assumed the quantum efficiency to be equal to 1). One of the most important property of the maximum signal-to-noise ratio following from Eq. (5.51) is its independence on the absorption coefficient as well as $\mu\tau$-product and other material parameters [230]. So the studied β-Ga_2O_3 detector is not inferior to the popular GaAs detectors in this theoretical limit.

The conventional holographic methods [231] employ the counting of the interference fringes in the pattern. The phase step between adjacent bright and dark fringes is of π, and this value is much higher than minimal detectable amplitude of our sensor. The better sensitivity of the adaptive detector mainly results from the narrow detection bandwidth, which is of order of 1 Hz in our setup and can easily be decreased to 0.01 Hz and lower in modern lock-in voltmeters. However, there is also a disadvantage of the developed technique with respect to the classical interferometry: we obtain the distribution of the phase with the local point-by-point measurements, while the traditional techniques allows the visualization of the whole phase distribution in the only experiment.

The point-by-point mapping of the phase amplitude implicitly utilizes the adaptive property of the detector. The wavefront of the light passed through the polished glass plate has no visible distortions, but the front and rear surfaces of a real plate are always wedged with a small angle. This means, that transversal movements of the plate would require the adjustment of the operating point in the conventional homodyne interferometer using a photodiode as the detector [250]. The dynamic space-charge gratings keep the optimal operating point of the non-steady-state photo-EMF detector and allow to avoid such adjustments. The heterodyne interferometers using photodiodes can also track the environmental phase shifts with the even better sensitivity than the discussed scheme, but their realization require the additional elements in the setup, namely the acousto-optical modulator and phase-demodulation circuit [250, 251]. Another advantage of the non-steady-state photo-EMF detector could reveal, if the plate had non-polished surfaces or was made of turbid media. We expect that the scattered light would also be effectively detected by the photo-EMF sensor, but such an experiment as well as more complicated data processing for the speckled light are considered as the tasks for the future work.

We have performed the experiment and simulation with the light polarized along y-axis, and the corresponding phase modulation amplitude Δ_y has been derived [Eq. (5.49)]. If we settled the polarization along x-axis keeping the applied force along y-axis, the analogous results would be obtained. The expression for the phase modulation amplitude Δ_x would be similar to Eq. (5.49) with σ_1 and σ_2 replaced by each other. The measurements of phases Δ_x and Δ_y for the known distribution of the stress field can be utilized for the determination of the piezooptic coefficients π_{11} and π_{12} in the isotropic media [246]. The more complicated problems, such as the measurement of all components of the stress tensor $\sigma_{kl}(x, y, z)$ or determination of the piezooptic coefficients π_{ijkl} in an anisotropic crystalline medium [252], will require the multi-directional illumination and solution of the inverse scattering problem typical for the optical tomography [253, 254].

References

[1] Gunter P.; Huignard J.-P. *Photorefractive effects and materials*. Topics in applied physics: Photorefractive materials and their applications Volume 61 (Eds. P. Gunter and J.-P. Huignard) Berlin, Heidelberg: Springer-Verlag, p. 7-73, 1988.

[2] Petrov M.P.; Stepanov S.I.; Khomenko A.V. *Photorefractive crystals in coherent optical systems;* Springer-Verlag: Berlin, 1991.

[3] Agullo-Lopez F.; Cabrera J.M.; Agullo-Rueda F. *Electrooptics: phenomena, materials and applications;* Academic Press: London, San Diego, New York, Boston, Sydney, Tokyo, Toronto, 1994.

[4] Nolte D.D. *Photorefractive transport and multi-wave mixing;* The Kluwer International Series in Engineering and Computer Science Photorefractive effects and materials (Edited by D. Nolte) Kluwer Academic Publishers: Boston, Dordrecht, London, 1995; pp 1-66.

[5] Solymar L.; Webb D.J.; Grunnet-Jepsen A. *The physics and applications of photorefractive materials;* Oxford: Clarendon Press, 1996.

[6] Deferrari H.A.; Darby R.A.; Andrews F.A. *J. Acost. Soc. Am.* 1967, 42, 982-990.

[7] Forward R.L. *Phys. Rev. D.* 1978, 17, 379-390.

[8] Gorecki C. *J. Optics (Paris)* 1995, 26, 29-34.

[9] Pernick B.J. *Appl. Opt.* 1973, 12, 607-610.

[10] Nokes M.A.; Hill B.C.; Barelli A.E. *Rev. Sci. Instrum.* 1978, 49, 722-728.

[11] Eberhardt F.J.; Andrews F.A. *J. Acoust. Soc. Am.* 1970, 48, 603-609.

[12] Dragsten P.R.; Webb W.W. *J. Acoust. Soc. Am.* 1976, 60, 665-671.

[13] Martinussen H.; Aksnes A.; Engan H. *Optics Express* 2007, 18, 11370-11384.

[14] Paul M.; Betz B.; Arnold W. *Appl. Phys. Lett.* 1987, 50, 1569-1571.

[15] Stepanov S.I.; Sokolov I.A. In *IEEE Proc. of Second International Conference on Holographic Systems, Components and Applications;* Institution of Electrical Engineers: London, U.K., 1989; Vol. 311, pp 95-100.

[16] Stepanov S.I., In *International Trends in Optics;* Goodman J.; Ed.; Academic Press: Boston, 1991; pp 124-140.

[17] Pouet B.; Krishnaswamy S. *Appl. Opt.* 1996, 35, 787-793.

[18] Stepanov S. In *Handbook of Advanced Electronic and Photonic Materials;* Nalwa H.S.; Ed.; Academic Press, 2000; Vol. 2, pp 205-272.

[19] Trofimov G.S.; Stepanov S.I. *Fiz. Tverd. Tela (Leningrad)* 1986, 28 2785-2789 [*Sov. Phys. Solid State* 1986, 28 1559-1562].

[20] Petrov M.P.; Sokolov I.A.; Stepanov S.I.; Trofimov G.S. *J. Appl. Phys.* 1990, 68, 2216-2225.

[21] Stepanov S.I.; Sokolov I.A.; Trofimov G.S.; Vlad V.I.; Popa D.; Apostol I. *Opt. Lett.* 1990, 15, 1239-1241.

[22] Sokolov I.A.; Bryushinin M.A. In *Advances in condensed matter and materials research;* Geelvinck H.; Reynst S.; Eds.; Nova Science Publishers, Inc.: New York, 2011; Vol. 10, pp 139-228.

[23] Sochava S.L.; Buse K.; Krätzig E. *Opt. Commun.* 1993, 105, 315-319.

[24] Sochava S.L.; Buse K.; Krätzig E. *Opt. Commun.* 1993, 98, 265-268.

[25] Gerwens A.; Buse K.; Krätzig E.; et al. *J. Opt. Soc. Am. B* 1998, 15, 2143-2146.

[26] Korneev N.; Mayorga D.; Stepanov S.; Gerwens A.; Buse K.; Kratzig E. *Opt. Commun.* 1998, 146, 215-219.

[27] Bittner R.; Meerholz K.; Stepanov S. *Appl. Phys. Lett.* 1999, 74, 3723-3725.

[28] Korneev N.; Mayorga D.; Stepanov S.; Veenhuis H.; Buse K.; Kuper C.; Hesse H.; Krätzig E. *Opt. Commun.* 1999, 160, 98-102.

[29] Seres I.; Stepanov S.; Mansurova S. *J. Opt. Soc. Am. B* 2000, 17, 1986 -1991.

[30] Trofimov G.S.; Kosarev A.I.; Kovrov A.G.; LeComber P.G. *J. Non-Cryst. Solids* 1991, 137,138, 483-486.

[31] Kosarev A.I.; Trofimov G.S. *Int. J. Electronics* 1994, 56, 1023-1028.

[32] Haken U.; Hundhausen M.; Ley L. *Phys. Rev. B* 1995, 51, 10579-10590.

[33] Pauliat G.; Roosen G. *J. Opt. Soc. Am. B* 1990, 7, 2259-2267.

[34] Strohkendl F.P. *J. Appl. Phys.* 1989, 65, 3773-3780.

[35] Hou S.L.; Lauer R.B.; Aldrich R.E. *J. Appl. Phys.* 1973, 44, 2652-2658.

[36] Kostyuk B.Kh.; Kudzin A.Yu.; Sokolyanskii G.Kh. *Sov. Phys. Solid State* 1980, 22, 1429-1432.

[37] Partanen J.P.; Jonathan J.M.C.; Hellwarth R.W. *Appl. Phys. Lett.* 1990, 57, 2404-2406.

[38] Sokolov I.A.; Stepanov S.I. *J. Opt. Soc. Am. B* 1993, 10, 1483-1488.

[39] Biaggio I.; Hellwarth R.W.; Partanen J.P. *Phys. Rev. Lett.* 1997, 78, 891-894.

[40] Bryushinin M.A.; Sokolov I.A. *Phys. Rev. B* 2000, 62, 7186-7194.

[41] Tayebati P.; Mahgerefteh D. *J. Opt. Soc. Am. B* 1991, 8, 1053-1064.

[42] Sokolov I.A.; Stepanov S.I. *Optik* 1993, 93, 175-182.

[43] G.C. Valley, *IEEE J. Quantum Electron.* 1983, QE-19, 1637-1645.

[44] Ryvkin S.M. *Photoelectric effects in semiconductors;* Consultants Bureau: New York, 1964.

[45] Refregier Ph.; Solymar L.; Rajbenbach H.; Huignard J.P. *J. Appl. Phys.* 1985, 58, 45-57.

[46] Bryushinin M.A.; Petrov A.A.; Sokolov I.A. *Appl. Phys. Lett.* 1999, 75, 445-447.

[47] Mokrushina E.V.; Bryushinin M.A.; Kulikov V.V.; Petrov A.A.; Sokolov I.A. *J. Opt. Soc. Am. B* 1999, 16, 57-62.

[48] Kazarinov R.F.; Suris R.A.; Fuks B.I. *Sov. Phys. Semicond.* 1973, 7, 102-107.

[49] Hutson A.R.; McFee J.H.; White D.L. *Phys. Rev. Lett.* 1961, 7, 237-239.

[50] Stepanov S.I.; Kulikov V.V.; Petrov M.P. *Opt. Commun.* 1982, 44, 19-23.

[51] Lyuksyutov S.F.; Buchhave P.; Vasnetsov M.V. *Phys. Rev. Lett.* 1997, 79, 67-70.

[52] Segev M.; Collings B.; Abraham D. *Phys. Rev. Lett.* 1996, 76, 3798-3801.

[53] Hundhausen M. *J. Non-Cryst. Solids* 1996, 198-200, 146-152.

[54] Kumar J.; Albanese G.; Steier W.H. *J. Opt. Soc. Am. B* 1987, 4, 1079-1082.

[55] Stepanov S.I.; Petrov M.P. *Opt. Commun.* 1985, 53, 292-295.

[56] Johansen P.M.; Pedersen H.C. *J. Opt. Soc. Am. B* 1998, 15, 1366-1374.

[57] Walsh K.; Powell A.K.; Stace C.; Hall T.J. *J. Opt. Soc. Am. B* 1990, 7, 288-303.

[58] Sokolov I.A.; Stepanov S.I. *Appl. Opt.* 1993, 31, 1958-1964.

[59] Sokolov I.A.; Stepanov S.I. *Electron. Lett.* 1990, 26, 1275-1277.

[60] Bryushinin M.; Kulikov V.; Sokolov I. *Phys. Rev. B* 2002, 65, 245204.

[61] Bryushinin M.A.; Sokolov I.A. *Phys. Rev. B* 2001, 63, 153203.

[62] van der Ziel A. *Noise. Sources, characterization, measurement;* Prentice-Hall. Inc.: Englewood Cliffs, N. J., 1970.

[63] Huignard J.P.; Marrakchi A. *Opt. Commun.* 1981, 38, 249-254.

[64] Hamel de Montchenault G.; Loiseaux B.; Huignard J.P. *Electron. Lett.* 1986, 22, 1030-1032.

[65] Heaton J.M.; Solymar L. *IEEE J. Quantum Electron.* 1988, 24, 558-567.

[66] Petrov M.P.; Petrov V.M.; Bryksin V.V.; Gerwens A.; Wevering S.; Krätzig E. *J. Opt. Soc. Am. B* 1998, 15, 1880-1888.

[67] Mallick S.; Imbert B.; Ducollet H.; Herriau J.P.; Huignard J.P. *J. Appl. Phys.* 1988, 63, 5660-5663.

[68] Pedersen H.C.; Johansen P.M. *Opt. Lett.* 1995, 20, 689-691.

[69] Petrov M.P.; Bryksin V.V.; Vogt H.; Rahe F.; Krätzig E. *Phys. Rev. B* 2002, 66, 085107.

[70] Bryushinin M. *Appl. Phys. B* 2004, 79, 851-856.

[71] Bryushinin M.A. *Zhurn. Tekh. Fiz. (St. Petersburg)* 2004, 74, 62-66 [*Tech. Phys.* 2004, 49, 1016-1020].

[72] Bryushinin M.; Kulikov V.; Sokolov I. *Phys. Rev. B* 2005, 71, 165208.

[73] Kogelnik H. *Bell. Syst. Tech. J.* 1969, 48, 2909-2947.

[74] Mansurova S.; Stepanov S.; Korneev N.; Dibon C. *Opt. Commun.* 1998, 152, 207-214.

[75] Sturman B.I.; Mann M.; Otten J.; Ringhofer K.H. *J. Opt. Soc. Am. B* 1993, 10, 1919-1932.

[76] Sochava S.L.; Buse K.; Krätzig E. *Phys. Rev. B* 1995, 51, 4684-4686.

[77] Greenaway D.L.; Nitsche R. *J. Phys. Chem. Solids* 1965, 26, 1445-1458.

[78] Domingo G.; Itoga R.S.; Kannewurf C.R. *Phys. Rev.* 1966, 143, 536-541.

[79] Au-Yang M.Y.; Cohen M.L. *Phys. Rev.* 1969, 178, 1279-1283.

[80] Patil S.G.; Tredgold R.H. *J. Phys. D* 1971, 4, 718-722.

[81] Nakata R.; Yamaguchi M.; Zembutsu S.; Sumita M. *J. Phys. Soc. Jap.* 1972, 32, 1153-1153.

[82] Shibata T.; Makanushi Y.; Miura T.; Kishi T. *J. Phys. Chem. Sol.* 1990, 51, 1297-1300.

[83] Dubrovskii G.B. *Fiz. Tverd. Tela (St. Petersburg)* 1998, 40, 1712-1718 [*Phys. Solid State* 1998, 40, 1557-1562].

[84] Bryushinin M.A.; Dubrovsky G.B.; Sokolov I.A. *Appl. Phys. B* 1999, 68, 871-875.

[85] Bryushinin M.A.; Dubrovskii G.B.; Petrov A.A.; Sokolov I.A.; Vainos N.A.; Kalpouzos C. *Fiz. Tverd. Tela (St. Petersburg)* 2002, 44, 1203-1205 [*Phys. Solid State* 2002, 44, 1254-1256].

[86] Stepanov S.I.; Trofimov G.S. *Fiz. Tverd. Tela (Leningrad)* 1989, 31, 89-92 [*Sov. Phys. Solid State* 1989, 31, 49-50].

[87] Korneev N.A.; Stepanov S.I. *J. Appl. Phys.* 1993, 74, 2736-2741.

[88] Korneev N.; Mansurova S.; Rodriguez P.; Stepanov S. *J. Opt. Soc. Am B* 1997, 14, 396-399.

[89] Landolt-Börnstein *Numerical data and functional relationships in science and technology;* Madelung O.; Ed.; Springer-Verlag: Berlin, 1983; Vol. 17, Subvol. f.

[90] Dubrovskii G.B.; Shelykh A.I. *Fiz. Tverd. Tela (St. Petersburg)* 1998, 40, 1426-1427 [*Phys. Solid State* 1998, 40, 1295-1296].

[91] Acharya S.; Srivastava O.N. *Phys. Stat. Sol. (a)* 1981, 65, 717-723.

[92] Chopra N.G.; Luyken R.J.; Cherrey K.; Crespi V.H.; Cohen M.L.; Louie S.G.; Zettl A. *Science* 1995, 269, 966-967.

[93] Sharupin B.N. Structure and properties of boron pyronitride. In *Chemical vapor deposition of refractory inorganic materials;* Shpak V.S.; Avarbe R.G.; Ed.; Leningrad, 1976.

[94] Geick R.; Perry C.H.; Rupprecht G. *Phys. Rev.* 1966, 146, 543-547.

[95] Zunger A.; Katzir A.; Halperin A. *Phys. Rev. B* 1976, 13, 5560-5573.

[96] Xu Y.-N.; Ching W.Y. *Phys. Rev. B* 1991, 44, 7787-7798.

[97] El-Yadouni A.; Soltani A.; Boudrioua A.; Thevenin P.; Bath A.; Loulergue J.C. *Opt. Mater.* 2001, 17, 319-322.

[98] Cappellini G.; Satta G.; Palummo M.; Onida G. *Phys. Rev. B* 2001, 64, 035104-1-11.

[99] Plaksin O.A.; Stepanov V.A.; Stepanov P.A.; Demenkov P.V.; Chernov V.M.; Krutskikh A.O., *Nucl. Inst. Met. Phys. Res. B* 2002, 193, 265-270.

[100] Sokolov I.A.; Bryushinin M.A.; Ordin S.V.; Kulikov V.V.; Petrov A.A. *J. Phys. D* 2006, 39, 1063-1068.

[101] Korneev N.A.; Stepanov S.I. *IEEE J. Quantum Electron.* 1994, 30, 2721-2725.

[102] Katzir A.; Suss J.T.; Zunger A.; Halperin A. *Phys. Rev. B* 1975, 11, 2370-2377.

[103] Grinyaev S.N.; Konusov F.V.; Lopatin V.V. *Phys. Solid State* 2002, 44, 286-293.

[104] Ilegems M. *J. Cryst. Growth* 1972, 13/14, 360-364.

[105] Crouch R.K.; Debnam W.J.; Fripp A.L. *J. Mater. Sci.* 1978, 13, 2358-2364.

[106] Kuznetsov N.I.; Nikolaev A.E.; Zubrilov A.S.; Melnik Yu.V.; Dmitriev V.A. *Appl. Phys. Lett.* 1999, 75, 3138-3140.

[107] Bryushinin M.; Golubev V.; Kumzerov Yu.; Kurdyukov D.; Sokolov I. *Physica B* 2009, 404 12511254.

[108] Bryushinin M.; Golubev V.; Kumzerov Yu.; Kurdyukov D.; Sokolov I. *Appl. Phys. B* 2009, 95 489495.

[109] Levitz P.; Ehret G.; Sinha S.K.; Drake J.M. *J. Chem. Phys.* 1991, 95, 6151-6161.

[110] Davydov V.Yu.; Dunin-Borkovski R.E.; Golubev V.G.; Hutchison J.L.; Kartenko N.F.; Kurdyukov D.A.; Pevtsov A.B.; Sharenkova N.V.; Sloan J.; Sorokin L.M. *Semicond. Sci. Technol.* 2001, 16, L5.

[111] Golosovsky I.V.; Delaplane R.G.; Naberezhnov A.A.; Kumzerov Y.A. *Phys. Rev. B* 2004, 69, 132301.

[112] Mott N.F.; Davis E.A. *Electron processes in non-crystalline materials;* Clarendon Press: Oxford, 1979.

[113] Lu W.; Lieber C. *J. Phys. D: Appl. Phys.* 2006, 39, R387-R406.

[114] Likharev K.K. *J. Nanoelectron. Optoelectron.* 2008, 3, 203-230.

[115] Li Y.; Qian F.; Xiang J.; Lieber C.M. *Mater. Today* 2006, 9, 18-27.

[116] Pauzauskie P.J.; Yang P. *Mater. Today* 2006, 9, 36-45.

[117] Duan X.; Huang Y.; Cui Y.; Wang J.; Lieber C.M. *Nature* 2001, 409 66-69.

[118] Kumzerov Y.; Vakhrushev S. *in Encyclopedia of nanoscience and nanotechnology;* Nalva H.S.; Ed.; American Scientific Publishers: Los Angeles 2004; Vol. 7, pp 811-849.

[119] Gates B.; Mayers B.; Cattle B.; Xia Y. *Adv. Funct. Mater.* 2002, 12, 219-227.

[120] Xiong S.; Xi B.; Wang W.; Wang C.; Fei L.; Zhou H.; Qian Y. *Cryst. Growth Des.* 2006, 6, 1711-1716.

[121] Cheng L.; Shao M.; Chen D.; Wei X.; Wang F.; Hua J. *J. Mater. Sci.: Mater. Electron.* 2008, 19, 1209-1213.

[122] Lu Q.; Gao F.; Komarneni S. *Chem. Mater.* 2006, 18, 159-163.

[123] Cao X.; Xie Y.; Zhang S.; Li F.; *Adv. Mater.* 2004, 16, 649-653.

[124] Gates B.; Mayers B.; Grossman A.; Xia Y. *Adv. Mater.* 2002, 14 1749-1752.

[125] Bogomolov V.N.; Kholodkevich S.V.; Romanov S.G.; Agroskin L.S. *Solid State Commun.* 1983, 47, 181-182.

[126] Bragg L.; Claringbull G.F. *Crystal structure of minerals;* G. Bell and Sons: London, 1965.

[127] Sobolev N.D. *Introduction to asbestos science (in Russian);* Nedra: Moscow, 1971.

[128] Qin D.; Tao H.; Zhao Y.; Lan L.; Chan K.; Cao Y. *Nanotechnology* 2008, 19, 355201.

[129] Sokolov I.A.; Stepanov S.I.; Trofimov G.S. *Zhurn. Tekhn. Fiz. (Leningrad)* 1989, 59, 126-129 [*Sov. Tech. Phys.* 1989, 34, 1165-1167].

[130] Bryushinin M.A.; Kulikov V.V.; Sokolov I.A. *Zhurn. Tekhn. Fiz. (St. Petersburg)* 2002, 72, 79-86 [*Tech. Phys.* 2002, 47, 1283-1290].

[131] Kukhtarev N.V.; Kukhtareva T.; Lyuksyutov S.F.; Reagan M.A.; Banerjee P.P.; Buchhave P. *J. Opt. Soc. Am. B* 2005, 22, 1917-1922.

[132] Veenhuis H.; Buse K.; Krätzig E.; Korneev N.; Mayorga D. *J. Appl. Phys.* 1999, 86, 2389-2392.

[133] Gather M.C.; Mansurova S.; Meerholz K. *Phys. Rev. B* 2007, 75, 165203.

[134] dos Santos T.O.; Frejlich J.; Shcherbin K. *Appl. Phys. B* 2010, 99, 701-707.

[135] Sokolov I.A.; Hess P.; Bryushinin M.A.; Kulikov V.V.; Khan S.H.; Grattan K.T.V. *in Interferometry in speckle light: Theory and applications;* Jacquot P.; Fournier J.-M.; Eds.; Springer: Berlin, Heidelberg 2000; pp. 187-194.

[136] Bryushinin M.A.; Grattan K.T.V.; Kulikov V.V.; Sokolov I.A. *J. Mod. Opt.* 2006, 53, 857-864.

[137] Sokolov I.A.; Bryushinin M.A.; Petrov A.A.; Zhilyaev Yu.V. *Nucl. Instr. Meth. Phys. Res. A* 2007, 571, 366-368.

[138] dos Santos T.O.; Frejlich J.; Launay J.C.; Shcherbin K. *Appl. Phys. B*, 2009, 95, 627-632.

[139] Mansurova S.; Moreno Zarate P.; Rodriguez P.; Stepanov S.; Köber S.; Meerholz K. *Opt. Lett.* 2012, 37, 383-385.

[140] Sokolov I.A.; Bryushinin M.A.; Kulikov V.V.; Khan S.H.; Grattan K.T.V. *Int. J. Lifecycle Perform. Eng.* 2013, 1, 270-277.

[141] Bryushinin M.; Kulikov V.; Sokolov I.; Delaye P.; Pauliat G. *EPL – Europhys. Lett.* 2014, 105, 64003.

[142] Bryushinin M.; Kulikov V.; Sokolov I.; Delaye P.; Pauliat G. *J. Opt. Soc. Am. B* 2014, 31, 723-729.

[143] Palmour J.W.; Edmond J.A.; Kong H.S.; Carter Jr. C.H. *Physica B* 1993, 185, 461-465.

[144] Morkoç H.; Strite S.; Gao G.B.; Lin M.E.; Sverdlov B.; Burns M. *J. Appl. Phys.* 1994, 76, 1363-1398.

[145] Casady J.B.; Johnson R.W. *Solid State Electron.* 1996, 39, 1409-1422.

[146] Brylinski C. *Diam. Relat. Mater.* 1997, 6, 1405-1413.

[147] Kukushkin S.A.; Osipov A.V.; Zhukov S.G.; Zavarin E.E.; Lundin W.V.; Sinitsyn M.A.; Rozhavskaya M.M.; Tsatsulnikov A.F.; Troshkov S.I.; Feoktistov N.A. *Tech. Phys. Lett.* 2012, 38, 297-299.

[148] Rogalla M.; Runge K.; Söldner-Rembold A. *Nucl. Phys. B – Proc. Sup.* 1999, 78, 516-520.

[149] Lebedev A.A.; Ivanov A.M.; Strokan N.B. *Semiconductors* 2004, 38, 125-147.

[150] Kalinina E.V. *Semiconductors* 2007, 41, 745-783.

[151] Petrov M.P.; Bryksin V.V.; Lebedev A.A.; Lemmer M.; Imlau M. *J. Appl. Phys.* 2005, 98, 083706.

[152] Lemmer M.; Hilling B.; Wöhlecke M.; Imlau M.; Lebedev A.A.; Bryksin V.V.; Petrov M.P. *Eur. Phys. J. B* 2007, 60, 9-14.

[153] Bryushinin M.A.; Sokolov I.A. *Tech. Phys. Lett.* 1999, 25, 366-368.

[154] Bryushinin M.; Kulikov V.; Mokhov E.; Nagalyuk S.; Sokolov I. *Phys. Rev. B* 2012, 86, 085209.

[155] Bryushinin M.; Kulikov V.; Sokolov I. *Phys. Rev. B* 2003, 67, 075202.

[156] Ryvkin S.M. *Photoelectric effects in semiconductors;* Consultants Bureau: New York 1964.

[157] Vodakov J.A.; Mokhov E.N. *US Patent 4147572* 1979.

[158] Kaminski P.; Kozlowski R.; Kozubal M.; Miczuga M.; Palczewska M.; Pawlowski M.G.; Pawlowski M. *Phys. Status Solidi C* 2007, 4, 2967-2971.

[159] Hundhausen M.; Püsche R.; Röhrl J.; Ley L. *Phys. Status Solidi B* 2008, 245, 1356-1368.

[160] Chen X.D.; Fung S.; Ling C.C.; Beling C.D.; Gong M. *J. Appl. Phys.* 2003, 94, 3004-3010.

[161] Taniyasu Y.; Kasu M.; Makimoto T. *Nature* 2006, 441, 325-328.

[162] Taylor K.M.; Lenie C. *J. Electrochem. Soc.* 1960, 107, 308-314.

[163] Vinogradov V.L.; Kostanovskii A.V.; Kirillin A.V. *High Temp. – High Press.* 1991, 23, 685-688.

[164] Slack G.A.; Tanzilli R.A.; Pohl R.O.; Vandersande J.W. *J. Phys. Chem. Solids* 1987, 48, 641-647.

[165] Chin V.W.L.; Tansley T.L.; Osotchan T. *J. Appl. Phys.* 1994, 75, 7365-7372.

[166] Nagai T.; Fujimura R.; Shimura T.; Kuroda K. *Opt. Lett.* 2010, 35, 2136-2138.

[167] Mokhov E.N.; Avdeev O.V.; Barash I.S.; Chemekova T.Yu.; Roenkov A.D.; Segal A.S.; Wolfson A.A.; Makarov Yu.N.; Ramm M.G.; Helava H. *J. Cryst. Growth* 2005, 281, 93-100.

[168] Collins A.T.; Lightowlers E.C.; Dean P.J. *Phys. Rev.* 1967, 158, 833-838.

[169] Moore W.J.; Freitas J.A.; Holm R.T.; Kovalenkov O.; Dmitriev V. *Appl. Phys. Lett.* 2005, 86, 141912.

[170] Edwards J.; Kawabe K.; Stevens G.; Tredgold R.H. *Solid State Commun.* 1965, 3, 99-100.

[171] Francis R.W.; Worrell W.L. *J. Electrochem. Soc.* 1976, 123, 430-433.

[172] Taniyasu Y.; Kasu M.; Makimoto T. *Appl. Phys. Lett.* 2004, 85, 4672-4674.

[173] Bryushinin M.A.; Kulikov V.V.; Mokrushina E.V.; Mokhov E.N.; Petrov A.A.; Sokolov I.A. *J. Phys. D: Appl. Phys.* 2014, 47, 415102.

[174] Bryushinin M.A.; Petrov A.A.; Pisarev R.V.; Sokolov I.A. *Phys. Solid State* 2015, 57, 907-913.

[175] Bryushinin M.A.; Sokolov I.A.; Pisarev R.V.; Balbashov A.M. *Opt. Express* 2015, 23, 32736-32746.

[176] Feng P.; Zhang J.Y.; Li Q.H.; Wang T.H. *Appl. Phys. Lett.* 2006, 88, 153107.

[177] Li L.; Auer E.; Liao M.; Fang X.; Zhai T.; Gautam U.K.; Lugstein A.; Koide Y.; Bandoa Y.; Golberga D. *Nanoscale* 2011, 3, 1120.

[178] Kokubun Y.; Miura K.; Endo F.; Nakagomi S. *Appl. Phys. Lett.* 2007, 90, 031912.

[179] Guo D.; Wu Z.; Li P.; An Y.; Liu H.; Guo X.; Yan H.; Wang G.; Sun C.; Li L.; Tang W. *Opt. Mater. Express* 2014, 4, 1067-1076.

[180] Oshima T.; Okuno T.; Arai N.; Suzuki N.; Ohira S.; Fujita S. *Appl. Phys. Express* 2008, 1, 011202.

[181] Orita M.; Ohta H.; Hirano M.; Hosono H. *Appl. Phys. Lett.* 2000, 77, 4166.

[182] Passlack M.; Schubert E.F.; Hobson W.S.; Hong M.; Moriya N.; Chu S.N.G.; Konstadinidis K.; Mannaerts J.P.; Schnoes M.L.; Zydzik G.J. *J. Appl. Phys.* 1995, 77, 686-693.

[183] Guo D.Y.; Wu Z.P.; An Y.H.; Li P.G.; Wang P.C.; Chu X.L.; Guo X.C.; Zhi Y.S.; Lei M.; Li L.H.; Tang W.H. *Appl. Phys. Lett.* 2015, 106, 042105.

[184] Oshima T.; Kaminaga K.; Mashiko H.; Mukai A.; Sasaki K.; Masui T.; Kuramata A.; Yamakoshi S.; Ohtomo A. *Jpn. J. Appl. Phys.* 2013, 52, 111102.

[185] Chang P.-C.; Fan Z.; Tseng W.-Y.; Rajagopal A.; Lu J.G. *Appl. Phys. Lett.* 2005, 87, 222102.

[186] Higashiwaki M.; Sasaki K.; Kuramata A.; Masui T.; Yamakoshi S. *Appl. Phys. Lett.* 2012, 100, 013504.

[187] Fleischer M.; Meixner H. *Sens. Actuators B* 1991, 5, 115-119.

[188] Geller S. *J. Chem. Phys.* 1960, 33, 676-684.

[189] Peelaers H.; Van de Walle C.G. *Phys. Status Solidi B* 2015, 252, 828-832.

[190] Liu B.; Gu M.; Liu X. *Appl. Phys. Lett.* 2007, 91, 172102.

[191] Balbashov A.M.; Egorov S.K. *J. Cryst. Growth*, 1981, 52, 498-504.

[192] Voronchikhina M.E.; Balbashov A.M.; Galagan B.I.; Shulman I.L. *in Scientific Program and Book of Abstracts: 17th International Conference on Crystal Growth and Epitaxy;* Warsaw, Poland, 2013, pp. 379-380.

[193] Ueda N.; Hosono H.; Waseda R.; Kawazoe H. *Appl. Phys. Lett.* 1997, 71, 933-935.

[194] Víllora E.G.; Shimamura K.; Yoshikawa Y.; Aoki K.; Ichinose N. *J. Cryst. Growth* 2004, 270, 420-426.

[195] Zhang J.; Li B.; Xia C.; Pei G.; Deng Q.; Yang Z.; Xu W.; Shi H.; Wu F.; Wu Y.; Xu J. *J. Phys. Chem. Solids* 2006, 67, 2448-2451.

[196] Bhaumik I.; Bhatt R.; Ganesamoorthy S.; Saxena A.; Karnal A.K.; Gupta P.K.; Sinha A.K.; Deb S.K. *Appl. Opt.* 2011, 50, 6006-6010.

[197] Lorenz M.R.; Woods J.F.; Gambino R.J. *J. Phys. Chem. Solids* 1967, 28, 403-404.

[198] Günter P.; Huignard J.-P. *Photorefractive materials and their applications II;* Springer: Berlin, Heidelberg, 1989.

[199] Ing R.K.; Monchalin J.-P. *Appl. Phys. Lett.* 1991, 59, 3233-3235.

[200] Wang C.C.; Davidson F.; Trivedi S. *Appl. Opt.* 1995, 34, 6496-6499.

[201] Wang C.C.; Linke R.A.; Nolte D.D.; Melloch M.R.; Trivedi S. *Appl. Phys. Lett.* 1997, 70, 2034-2036.

[202] Pepper D.M.; Dunning G.J.; Nolte D.D.; Coy J.A.; Pouet B.; Bacher G.D.; Klein M.B. *Opt. Photon. News* 1999, 10, 11-12.

[203] Delaye Ph.; de Rossi S.; Roosen G. *J. Opt. A: Pure Appl. Opt.* 2000, 2, 209-215.

[204] Kamshilin A.A.; Romashko R.V.; Kulchin Y.N. *J. Appl. Phys.* 2009, 105, 031101.

[205] Kolegov A.A.; Shandarov S.M.; Simonova G.V.; Kabanova L.A.; Burimov N.I.; Shmakov S.S.; Bykov V.I.; Kargin Yu.F. *Quantum Electron.* 2011, 41, 847-852.

[206] Valley G.C. *J. Opt. Soc. Am. B* 1984, 1, 868-873.

[207] Kamshilin A.A.; Frejlich J.; Cescato L. *Appl. Opt.* 1986, 25, 2375-2381.

[208] Bian S.; Frejlich J. *J. Mod. Opt.* 1996, 43, 1185-1198.

[209] de Oliveira I.; Freschi A.A.; Fier I.; Frejlich J. *Opt. Mater. Express* 2012, 2, 228-234.

[210] Trofimov G.S.; Stepanov S.I.; Petrov M.P.; Krasin'kova M.V. textitSov. Tech. Phys. Lett. 1987, 13, 108-109.

[211] Korneev N.; Mansurova S.; Stepanov S. *J. Appl. Phys.* 1995, 78, 2925-2931.

[212] Pauliat G.; Roosen G. *Ferroelectrics* 1987, 75, 281-294.

[213] Plesovskikh A.M.; Shandarov S.M.; Mart'yanov A.G.; Mandel' A.E.; Burimov N.I.; Shaganova E.A.; Kargin Yu.F.; Volkov V.V.; Egorysheva A.V. *Quantum Electron.* 2005, 35, 163-168.

[214] Romashko R.V.; Di Girolamo S.; Kulchin Y.N.; Kamshilin A.A. *J. Opt. Soc. Am. B* 2010, 27, 311-317.

[215] Collier R.J.; Burckhardt C.B.; Lin L.H. *Optical holography;* Academic Press: New York, 1971.

[216] Jones R.; Wykes C. *Holographic and speckle interferometry;* Cambridge University Press: Cambridge, 1989.

[217] Freschi A.A.; Frejlich J. *J. Opt. Soc. Am. B* 1994, 11, 1837-1841.

[218] Garcia P.M.; Buse K.; Kip D.; Frejlich J. *Opt. Commun.* 1995, 117, 235-240.

[219] Kamenov V.P.; Ringhofer K.H.; Sturman B.I.; Frejlich J. *Phys. Rev. A* 1997, 56, R2541-R2544.

[220] Podivilov E.V.; Sturman B.I.; Odoulov S.G.; Pavlyuk S.; Shcherbin K.V.; Gayvoronsky V.Ya.; Ringhofer K.H.; Kamenov V.P. *Phys. Rev. A* 2001, 63, 053805.

[221] Garcia P.M.; Freschi A.A.; Frejlich J.; Krätzig E. *Appl. Phys. B* 1996, 63, 207-208.

[222] Gorkunov M.; Sturman B.; Luennemann M.; Buse K. *Appl. Phys. B* 2003, 77, 43-48.

[223] Sturman B.; Podivilov E.; Gorkunov M. *Phys. Rev. E* 2005, 72, 016621.

[224] Rabinovich M.I.; Trubetskov D.I. *Oscillations and waves in linear and nonlinear systems;* Kluwer Academic Publishers, 1989.

[225] Luenberger D.G. *Introduction to dynamic systems;* John Wiley & Sons: New York, 1979.

[226] Bechhoefer J. *Rev. Mod. Phys.* 2005, 77, 783-836.

[227] Collins S.A.; Wasmundt K.C. *Opt. Eng.* 1980, 19, 478-487.

[228] Bryksin V.V.; Kleinert P.; Petrov M.P. *Phys. Solid State* 2004, 46, 1613-1621.

[229] McClelland T.E.; Webb D.J.; Sturman B.I.; Shamonina E.; Mann M.; Ringhofer K.H. *Opt. Commun.* 1996, 131, 315-321.

[230] Stepanov S.I. *Appl. Opt.* 1994, 33, 915-920.

[231] Vest C.M. *Holographic interferometry*. New York: John Wiley & Sons, 1979.

[232] Yu P.; Peng L.; Nolte D.D.; Melloch M.R. *Opt. Lett.* 2003, 28, 819-821.

[233] Ramaz F.; Forget B.C.; Atlan M.; Boccara A.C.; Gross M.; Delaye P.; Roosen G. *Opt. Express* 2004, 12, 5469-5474.

[234] Murray T.W.; Sui L.; Maguluri G.; Roy R.A.; Nieva A.; Blonigen F.; DiMarzio C.A. *Opt. Lett.* 2004, 29, 2509-2511.

[235] Rousseau G.; Blouin A.; Monchalin J.-P. *Opt. Express* 2008, 16, 12577-12590.

[236] Shandarov S.M.; Shmakov S.S.; Burimov N.I.; Syuvaeva O.S.; Kargin Y.F.; Petrov V.M. *JETP Lett.* 2012, 95, 618-621.

[237] Romashko R.V.; Efimov T.A.; Kulchin Y.N. *Quantum Electron.* 2014, 44, 269-273.

[238] Romashko R.V.; Kulchin Y.N.; Nippolainen E. *Laser Phys.* 2014, 24, 115604.

[239] Sokolov I.A.; Bryushinin M.A. In *Handbook of interferometers: Research, technology and applications;* Halsey D.; Raynor W.; Eds.; Nova Science Publishers, Inc.: New York, 2009; pp 669-689.

[240] Bryushinin M.A.; Kulikov V.V.; Mokhov E.N.; Romashko R.V.; Sokolov I.A. *Optik* 2016, 127, 341-344.

[241] Bryushinin M.A.; Sokolov I.A.; Zavestovskaya I.N.; Kulchin Yu.N. *B. Lebedev Phys. Inst.* 2016, 43, 35-36.

[242] Bryushinin M.A.; Sokolov I.A.; Balbashov A.M. *Appl. Opt.* 2017, 52, 170-178.

[243] Razeghi M.; Rogalski A. *J. Appl. Phys.* 1996, 79, 7433-7473.

[244] Sirotin Yu.I.; Shaskol'skaya M.P. *Fundamentals of crystal physics;* Nauka: Moscow, 1975 (in Russian).

[245] Landau L.D.; Pitaevskii L.P.; Kosevich A.M.; Lifshitz E.M. *Theory of elasticity;* Elsevier: Oxford, 1986.

[246] Krupych O.; Savaryn V.; Skab I.; Vlokh R. *Ukr. J. Phys. Opt.* 2011, 12, 150-159.

[247] Hutchinson J.R. *J. Acoust. Soc. Am.* 1972, 51, 233-240.

[248] Hutchinson J.R. *J. Appl. Mech.* 1980, 47, 901-907.

[249] Ebenezer D.D.; Ravichandran K.; Padmanabhan C. *J. Sound Vib.* 2005, 282, 991-1007.

[250] Dewhurst R.J.; Shan Q. *Meas. Sci. Technol.* 1999, 10, R139-R168.

[251] Eberhardt F.J.; Andrews F.A. *J. Acoust. Soc. Am.* 1970, 48, 603-609.

[252] Krupych O.; Savaryn V.; Vlokh R. *Appl. Opt.* 2014, 53, B1-B7.

[253] Sweeney D.W.; Vest C.M. *Appl. Opt.* 1973, 12, 2649-2664.

[254] Iwata K.; Nagata R. *Jpn. J. Appl. Phys.* 1975, 14, 379-383.

uthor Contact Information

Dr. Igor A. Sokolov
Leading Scientist
Ioffe Institute
194021 Polytekhnicheskaya 26, St.-Petersburg, Russia
Email: i.a.sokolov@mail.ioffe.ru; i.a.sokolov1963@gmail.com

Dr. Mikhail A. Bryushinin
Senior Scientist
Ioffe Institute
194021 Polytekhnicheskaya 26, St.-Petersburg, Russia
Email: mb@mail.ioffe.ru

Index

A

B

C

D

E

F

G

P

Q

R

S

T

U

V

W

X